Discovering
Advanced
Algebra
THIRD EDITION

More Practice Your Skills
Student Workbook

Discovering
Mathematics

Kendall Hunt
publishing company

Cover image © Shutterstock, Inc.

www.kendallhunt.com
Send all inquiries to:
4050 Westmark Drive
Dubuque, IA 52004-1840
1-800-542-6657

Copyright © 2002, 2007 by Key Curriculum Press
Copyright © 2017 by Kendall Hunt Publishing Company

ISBN 978-1-4652-9042-7

All rights reserved. No part of this publication may be reproduced, stored in a retrieval system, or transmitted, in any form or by any means, electronic, mechanical, photocopying, recording, or otherwise, without the prior written permission of the copyright owner.

Published in the United States of America

Contents

Developing Mathematical Thinking	1
Linear Modeling	5
Systems of Equations and Inequalities	15
Functions and Relations	21
Exponential, Power, and Logarithmic Functions	29
Quadratic Functions and Relations	37
Polynomial and Rational Functions	47
Trigonometry and Trigonometric Functions	57
Probability	65
Applications of Statistics	71

Developing Mathematical Thinking

Modeling and Tools

Name _____ Period _____ Date _____

1. Find the slope of each line.

   a. 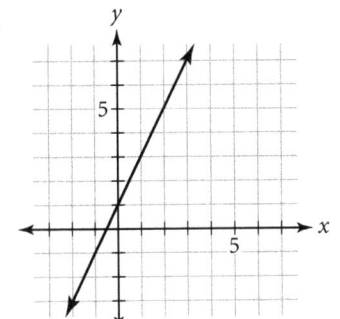   b. 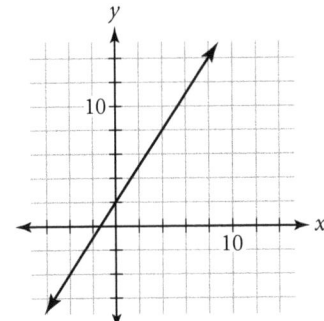   c.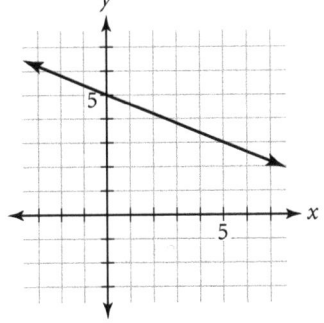

2. Use a coordinate graph to find the slope of the line that passes through each pair of points.

   a. $(0, 1)$ and $(3, 3)$ b. $(-5, -2)$ and $(-8, 4)$

   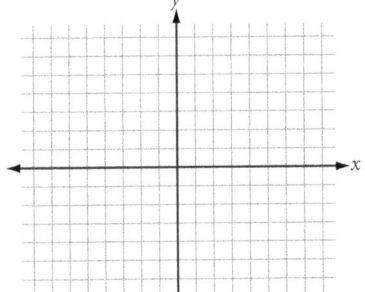   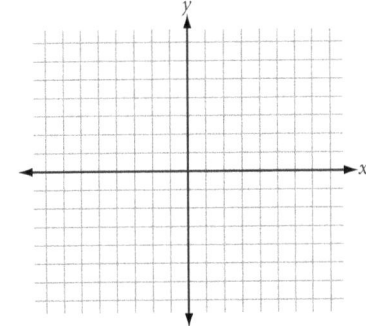

3. Describe a model that could be used to solve each of the following problems. Then solve the problem.

   a. A snail is at the bottom of a 12 foot well. Every day the snail crawls up 3 feet, but then slips down 2 feet each night. How many days will it take the snail to get to the top of the well?

   b. Three adults and 2 kids need to cross a river in a small row boat. The boat can carry two kids or one adult. How many times will the boat cross the river to get everyone to the other side?

   c. If six people met at a party and all shook hands with each other, how many handshakes would be exchanged?

Discovering Advanced Algebra More Practice Your Skills

Reasoning and Explaining

Name _____ Period _____ Date _____

1. An elevator in the Empire State Building is stopped at a floor in the middle of the building. From there it ascends up to the rooftop deck at a constant rate of speed. Using x to represent time and y to represent height from the ground, which equation best represents the height of the elevator as it ascends at a steady rate?

   A. $y = 50 + 0.725x$
   B. $y = 50 - 0.725x$
   C. $y = -50 - 0.725x$
   D. $y = -50 + 0.725x$

2. The company "DVD2U" decides to set up a membership club to make DVD's cheaper for their customers. Membership will cost $6.87 and for members the cost for each DVD will be $4.98. What equation best represents the total cost for buying DVD's from "DVD2U" if you are a member?

   A. $y = 4.98 + 6.87x$
   B. $y = 11.85 + x$
   C. $y = 6.87 + 4.98x$
   D. $y = 6.87(x - 4.98) + 11.85$

3. The perimeter of a rectangular playing field is 504 yards. The area of the playing field is 14,276 square yards. Write a system of equations that could be used to find the values of l, the length of the field, and w, the width of the field, that will satisfy these conditions.

4. In a basketball game, Kristin made 16 baskets. Each of the baskets was worth either 2 points or 3 points, and Kristin scored a total of 39 points in the game. Let x represent the number of two-point baskets and y represent the number of three-point baskets she made. Write a system of equations in terms of x and y to model the situation.

5. Holly sold 40 tickets for a concert. She sold x tickets at $20 each and y tickets at $30 each. She collected $8800. Write two equations connecting x and y.

Looking for and Generalizing Patterns

Name _____ Period _____ Date _____

1. Use units to help you find the missing information.

   a. If you live 2.5 miles from school, how many inches is that?

   b. If 16 oz = 1 lb, how many ounces are in 3.4 lbs?

2. Midway through a 2000-meter race, a photo is taken of five runners. It shows Sue 20 meters behind Erica. Erica is 50 meters ahead of Linda, who is 20 meters behind Opal. Opal is 40 meters behind Ruth. Who is in last place? In your diagram, use *S* for Sue, *E* for Erica, and so on.

3. Alfonso and Joan sold sandwiches for $1.75 each and juice boxes for $1.10 each at a football game. They sold a total of 540 items and took in $730.50.

   a. Identify the unknown quantities and assign variables.

   b. Write a system of equations to represent the situation.

   c. Which of the ordered pairs (s, j) is a solution for the problem?

       i. (330, 210)   ii. (210, 330)   iii. (367.50, 363)   iv. (577.50, 210)

   d. Interpret the solution from 3c according to the context of the problem.

4. As part of their homework assignment, Jason and Jordyn each found equations from a table of data relating miles and kilometers. One entry in the table paired 150 kilometers and 93 miles. From this pair of data values, Jason and Jordyn wrote different equations.

   a. Jason wrote the equation $y = 1.61x$. How did he get it? What does 1.61 represent? What do x and y represent?

   b. Jordyn wrote $y = 0.62x$ as her equation. What does 0.62 represent? What do x and y represent?

   c. Whose equation would you use to convert miles to kilometers?

   d. When would you use the other student's equation?

Linear Modeling

Recursively Defined Sequences

Name _____ Period _____ Date _____

1. Find the common difference, d, for each arithmetic sequence and the common ratio, r, for each geometric sequence.

   a. $1.5, 1.0, 0.5, 0, -0.5, \ldots$ b. $0.0625, 0.125, 0.25, \ldots$ c. $-1, 0.2, -0.04, 0.008, \ldots$

2. Write the first six terms of each sequence and identify each sequence as arithmetic or geometric.

   a. $u_1 = -18$
      $u_n = u_{n-1} + 6$ where $n \geq 2$

   b. $u_1 = 0.5$
      $u_n = 3u_{n-1}$ where $n \geq 2$

3. Write a recursive formula to generate each sequence. Then find the indicated term.

   a. $17.25, 14.94, 12.63, 10.32, \ldots$ Find the 15th term.

   b. $-2, 4, -8, 16, \ldots$ Find the 15th term.

4. Indicate whether each situation could be represented by an arithmetic sequence or a geometric sequence. Give the value of the common difference, d, for each arithmetic sequence and of the common ratio, r, for each geometric sequence.

   a. Phil rented an apartment for $850 a month. Each time he renewed his annual lease over the next 3 years, his landlord raised the rent by $50.

   b. Leora was hired as a first-year teacher at an annual salary of $30,000. She received an annual salary increase of 5% for each of the next 4 years.

5. Write a recursive formula for the sequence graphed at right. Find the 42nd term.

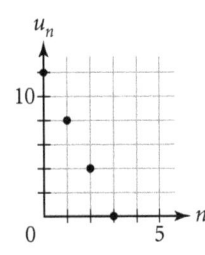

Discovering Advanced Algebra More Practice Your Skills
©2017 Kendall Hunt Publishing

Modeling Growth and Decay

Name _____ Period _____ Date _____

1. Find the common ratio for each sequence and identify the sequence as growth or decay. Give the percent increase or decrease for each.

   a. 42, 126, 378, 1134, . . .

   b. 19.2, 3.84, 0.768, 0.1536, . . .

   c. 90, 99, 108.9, 119.79, . . .

   d. 1800, 1080, 648, 388.8, . . .

2. Write a recursive formula for each sequence in Exercise 1. Use u_0 for the first term given and find u_5.

3. Factor each expression so that the variable appears only once. For example, $x + 0.05x$ factors into $x(1 + 0.05)$.

   a. $y - 0.19y$
   b. $2A - 0.33A$
   c. $u_{n-1} - 0.72u_{n-1}$
   d. $3u_{n-1} - 0.5u_{n-1}$

4. Write a recursive formula for the sequence 3, −8.5, 26, −77.5, ...

5. Match each recursive formula to a graph.

   a. $u_0 = 35$
      $u_n = (1 - 0.3) \cdot u_{n-1}$ where $n \geq 1$

   b. $u_0 = 35$
      $u_n = (1 - 0.5) \cdot u_{n-1}$ where $n \geq 1$

   c. $u_0 = 35$
      $u_n = -0.5 + u_{n-1}$ where $n \geq 1$

i.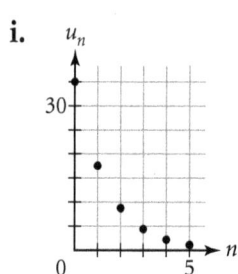

ii.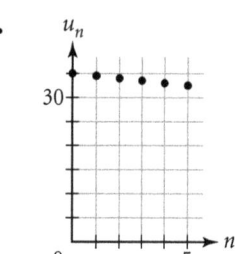

iii.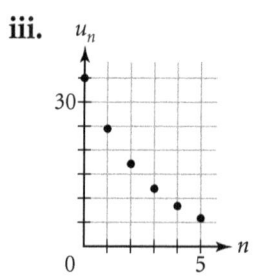

Linear Modeling
©2017 Kendall Hunt Publishing

Applications and Other Sequences

Name _____ Period _____ Date _____

1. For each sequence, find the value of u_1, u_2, and u_3. Identify the type of sequence (arithmetic, geometric, or shifted geometric) and tell whether it is increasing or decreasing.

a. $u_0 = 25$
$u_n = u_{n-1} + 8$ where $n \geq 1$

b. $u_0 = 10$
$u_n = 0.1 u_{n-1}$ where $n \geq 1$

c. $u_0 = 48$
$u_n = u_{n-1} - 6.9$ where $n \geq 1$

d. $u_0 = 500$
$u_n = (1 - 0.80) u_{n-1} + 25$ where $n \geq 1$

2. Solve.

a. $r = 0.9r + 30$

b. $s = 25 + 0.75s$

c. $t = 0.82t$

d. $v = 45 + v$

e. $w = 0.60w - 20$

f. $z = 0.125z + 49$

3. Find the long-run value for each sequence.

a. $u_0 = 48$
$u_n = 0.75 u_{n-1} + 25$ where $n \geq 1$

b. $u_0 = 12$
$u_n = 0.9 u_{n-1} + 2$ where $n \geq 1$

c. $u_0 = 62$
$u_n = (1 - 0.2) u_{n-1}$ where $n \geq 1$

d. $u_0 = 45$
$u_n = (1 - 0.05) u_{n-1} + 5$ where $n \geq 1$

4. Write a recursive formula for each sequence. Use u_0 for the first term given.

a. 0, 20, 36, 48.8, . . .

b. 100, 160, 226, 298.6, . . .

c. 50, 36, 27.6, 22.56, . . .

d. 40, 44, 50.4, 60.64, . . .

Discovering Advanced Algebra More Practice Your Skills
©2017 Kendall Hunt Publishing

Graphing Sequences

Name _____ Period _____ Date _____

1. Write five ordered pairs that represent points on the graph of each sequence.

   a. $b_0 = 2$
      $b_n = b_{n-1} + 8$ where $n \geq 1$

   b. $b_0 = 10$
      $b_n = 0.1 b_{n-1}$ where $n \geq 1$

   c. $b_0 = 0$
      $b_n = 2.5 b_{n-1} + 10$ where $n \geq 1$

   d. $b_0 = 150$
      $b_n = 0.8 b_{n-1} - 10$ where $n \geq 1$

2. Match each formula with a graph and identify the sequence as arithmetic or geometric.

   a. $u_0 = 10$
      $u_n = 1.5 u_{n-1}$ where $n \geq 1$

   b. $u_0 = 30$
      $u_n = u_{n-1} + 5$ where $n \geq 1$

   c. $u_0 = 80$
      $u_n = 0.75 u_{n-1}$ where $n \geq 1$

   i.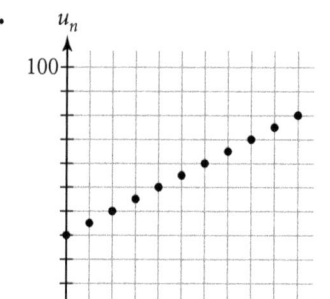

   ii.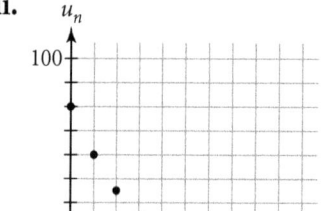

   iii.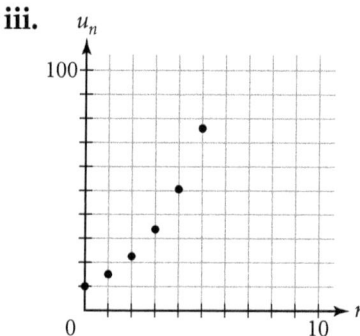

3. Imagine the graphs of the sequences generated by these recursive formulas. Describe each graph using exactly three of these terms: arithmetic, geometric, shifted geometric, linear, nonlinear, increasing, decreasing.

   a. $t_0 = 50$
      $t_n = t_{n-1} - 10$ where $n \geq 1$

   b. $a_0 = 1000$
      $a_n = 0.7 a_{n-1} + 100$ where $n \geq 1$

   c. $u_0 = 35$
      $u_n = u_{n-1} \cdot 1.75$ where $n \geq 1$

   d. $t_0 = 150$
      $t_n = (1 - 0.15) t_{n-1}$ where $n \geq 1$

Linear Equations and Arithmetic Sequences

Name _____ Period _____ Date _____

1. Find an explicit formula for each recursively defined arithmetic sequence.

   a. $u_0 = 18.25$
      $u_n = u_{n-1} - 4.75$ where $n \geq 1$

   b. $t_0 = 0$
      $t_n = t_{n-1} + 100$ where $n \geq 1$

2. Refer to the graph of the sequence.

   a. Write a recursive formula for the sequence. What is the common difference? What is the value of u_0?

   b. What is the slope of the line through the points? What is the y-intercept?

   c. Write an equation for the line that contains these points.

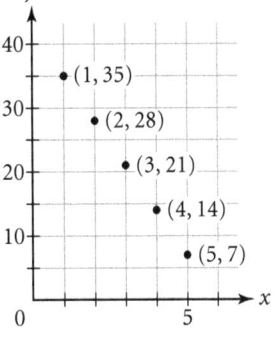

3. Find the slope of each line.

   a. $y = 5 + 3x$ b. $y = 10 - x$ c. $y = 0.6x - 0.8$

4. Write an equation in the form $y = a + bx$ for each line.

   a. The line that passes through the points of an arithmetic sequence with $u_0 = 11$ and a common difference of 9

   b. The line that passes through the points of an arithmetic sequence with $u_0 = -7.5$ and a common difference of -12.5

Modeling with Intercept and Slope

Name _____   Period _____   Date _____

1. Find the slope of the line containing each pair of points.

   a. (2, 6) and (4, 12)

   b. (0, 7) and (5, 0)

   c. $\left(\frac{1}{3}, \frac{2}{3}\right)$ and $\left(\frac{5}{6}, -\frac{1}{6}\right)$

2. Find the slope of each line.

   a. $y = 1.6 - 2.5x$

   b. $y = -4(x - 7) + 12$

   c. $y = 14.5 - 0.3(x - 30)$

3. Solve.

   a. $y = 6 - 2x$ for y if $x = -4$

   b. $y = a - 0.4x$ for a if $x = 600$ and $y = 150$

4. Find the equations of both lines in each graph.

   a.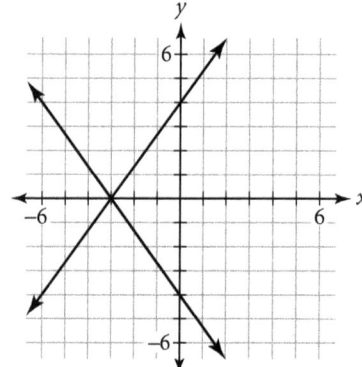

   b.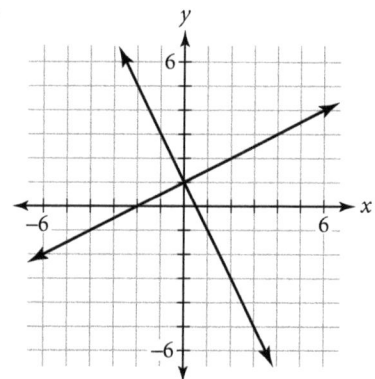

   c. What do the equations in part a have in common? What do you notice about their graphs?

   d. What do the equations in part b have in common? What do you notice about their graphs?

Linear Modeling
©2017 Kendall Hunt Publishing

Models and Predictions

Name _____ Period _____ Date _____

1. Write an equation in point-slope form for each line.

   a.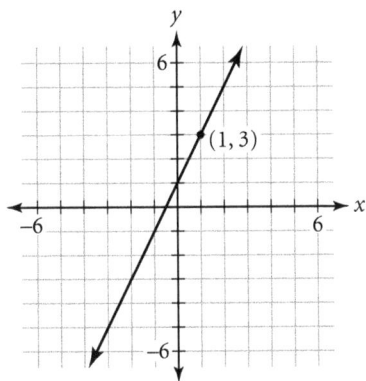

   b.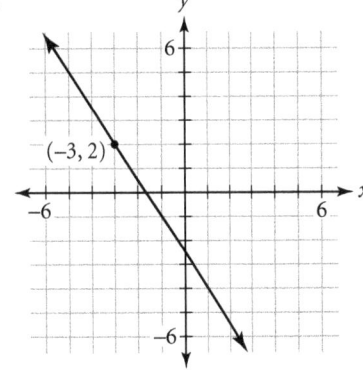

2. Write an equation in point-slope form for each line.

   a. Slope 0.75 and passing through $(-4, 10)$

   b. Parallel to $y = 7 - 4x$ and passing through $(2, -5)$

3. Solve.

   a. $d = 9 - 4(t + 5)$ for d if $t = 20$.

   b. $y = 500 - 20(x - 5)$ for x if $y = 240$.

   c. $a_n = -3.5 + 0.4(n - 12)$ for n if $a_n = 2.9$.

4. For each graph, use your ruler to draw a line of fit. Explain how your line satisfies the guidelines in your book.

   a.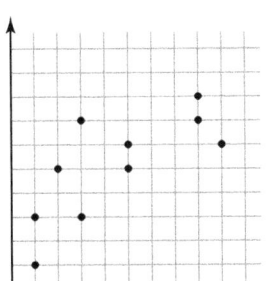

   b.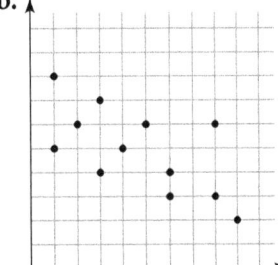

Residuals and Fit

Name _____ Period _____ Date _____

1. Determine whether the given point lies above or below the given line.

   a. $y = -2x + 6$; $(3, 1)$ b. $y = 3.6x - 18.8$; $(10, 16.9)$

2. The best fit line for a set of data is $\hat{y} = 4x - 5$. The table at right gives the x-value and the residual for each data point. Find the y-value for each data point.

x-value	0	1	3	10
Residual	1	-1	2	-3

3. This table gives the number of students enrolled in U.S. public schools for various years.

   a. Find the best fit line for the data. Round all answers to one decimal place. Does the y-intercept make sense for the data?

   b. Calculate the residuals.

School year	Public school enrollment
1909–10	17,814
1919–20	21,578
1929–30	25,678
1939–40	25,434
1949–50	25,111
1959–60	35,182
1969–70	45,550
1979–80	41,651
1989–90	40,543
1999–2000	46,812

   (*The World Almanac and Book of Facts 2007*)

   c. The *World Almanac* predicted that the public school enrollment in the 2013–14 school year would be 49,737 students. Use your best fit line to find the prediction, based on the data from the *Almanac*, for the enrollment in 2013–14 and calculate the residual of the *Almanac*'s prediction.

Least Squares Line

Name _____ Period _____ Date _____

1. Given the lists below, use your calculator to find the following.

List x	86	71	51	89	65	80	74
List y	97	69	55	109	75	96	84

   a. The mean and standard deviation of each list.

   b. The correlation and the equation of the least squares line.

   c. Reverse the two lists and find the correlation and the equation of the least squares line.

2. Given the lists below, use your calculator to find the following.

List x	9	62	6	59	34	73
List y	85	42	79	48	53	13

   a. The mean and standard deviation of each list.

   b. The correlation and the equation of the least squares line.

   c. Reverse the two lists and find the correlation and the equation of the least squares line.

Discovering Advanced Algebra More Practice Your Skills
©2017 Kendall Hunt Publishing

Systems of Equations and Inequalities

Linear Systems

Name _____ Period _____ Date _____

1. Identify the point of intersection listed below each system of linear equations that is the solution of that system.

a. $\begin{cases} 2x + 5y = 10 \\ x - 3y = -6 \end{cases}$
$(5, 0); (0, 2); (3, 1)$

b. $\begin{cases} 4x + 3y = 4 \\ 3x - 2y = -14 \end{cases}$
$(-2, 4); \left(0, \frac{4}{3}\right); (0, 7)$

c. $\begin{cases} 6x - 5y = 0 \\ x - y = -1 \end{cases}$
$(0, 0); (-5, -6); (5, 6)$

2. Write a system of linear equations that has each ordered pair as its solution.

a. $(5, 4)$

b. $(-3, 8)$

c. $(3, 10.5)$

3. Write an equation for each line described.

a. Perpendicular to $y = 2x - 3$ and passing through the point $(5, -4)$

b. Perpendicular to $y = 1.5 + 0.25x$ and passing through the point $(5, -2)$

4. Solve.

a. $8 - 3(x - 2) = 5 + 6x$

b. $3.8t - 16.2 = 12 + 2.8(t + 3)$

5. Use substitution to find the point (x, y) where each pair of lines intersect. Use a graph or table to verify your answer.

a. $\begin{cases} y = 3 - 2x \\ y = 5 + 2x \end{cases}$

b. $\begin{cases} y = 0.45x - 2 \\ y = -0.45x + 2 \end{cases}$

c. $\begin{cases} y = 9 + 4(x - 3) \\ y = 15 - 2x \end{cases}$

Substitution and Elimination

Name _____ Period _____ Date _____

1. Solve each equation for the specified variable.

   a. $r - s = 20$, for s

   b. $5x - 8y = -10$, for x

   c. $0.2m - 0.5n = 1$, for n

   d. $250x + 400y = -50$, for y

2. Graph each system and find an approximate solution. Then choose a method and find the exact solution. List each solution as an ordered pair.

   a. $\begin{cases} x + y = 1 \\ 2x - 2y = 1 \end{cases}$

   b. $\begin{cases} 3x - 2y = 6 \\ -2x + 3y = 0 \end{cases}$

   c. $\begin{cases} 5x + 4y = 16 \\ 4x - 3y = 12 \end{cases}$

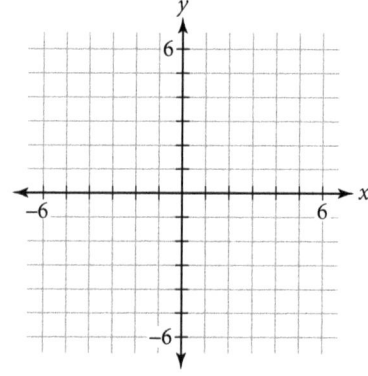
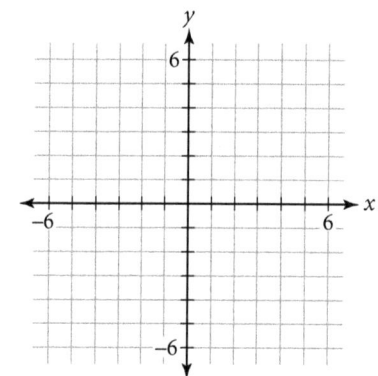
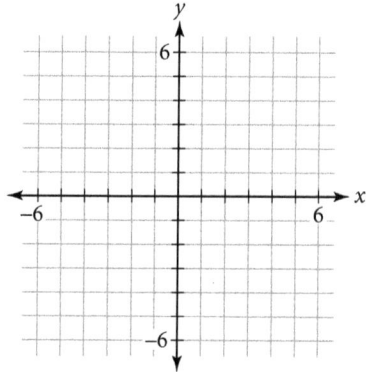

3. Solve each system of equations.

   a. $\begin{cases} 3x - 4y = 8 \\ y = x - 1 \end{cases}$

   b. $\begin{cases} 5x - 8y = 8 \\ -10x + 4y = -7 \end{cases}$

   c. $\begin{cases} 0.5x + 1.5y = 5 \\ x + y = -10 \end{cases}$

4. Classify each system as consistent or inconsistent. If a system is consistent, classify it as dependent or independent.

   a. $\begin{cases} -3x + 2y = 8 \\ y = 4 - x \end{cases}$

   b. $\begin{cases} 6m + 3n = 15 \\ n = -2m + 5 \end{cases}$

   c. $\begin{cases} k = 2j + 9 \\ 4j - 2k = 3 \end{cases}$

Linear and Non-Linear Systems of Equations

Name _____ Period _____ Date _____

1. Solve each equation for y.

   a. $5x - y = -10$ b. $xy + 3y = 9$ c. $14x + 7y = 42$

   d. $2xy = 12.6$ e. $2.2xy + 1.1y + 5.5 = 0$ f. $\dfrac{6 + 3x}{y + 2} = 3$

2. Complete each equation so that the point $(3, 4)$ is a solution.

   a. $3x + 4y = ?$ b. $?x + 6y = 30$ c. $4x + ?y = 14$

3. Determine whether or not the given point is a solution to the given system.

   a. $(9, 7)$
   $2x + 3y = 36$
   $xy = 54$

   b. $\left(3\sqrt{5}, 6\sqrt{5}\right)$
   $x^2 + y^2 = 225$
   $y = 2x$

   c. $(2, -2)$
   $4xy = 16$
   $\dfrac{4 + 2x}{y + 6} = 4$

4. Solve each system of equations.

   a. $xy = -12$
   $x - 2y = -14$

   b. $5x - 2y = 10$
   $x = \dfrac{60}{y}$

   c. $\dfrac{y + 5}{x - 2} = 5$
   $x^2 - x - 6 = y$

5. Solve each system of equations symbolically.

   a. $x^2 + y^2 = 5$
   $y = 2x - 5$

   b. $xy + 3y = 9$
   $2x - y = 1$

   c. $14x + 7y = 42$
   $y = x^2 + 2x + 1$

6. Solve each system of equations by graphing.

   a. $x^2 + y^2 = 25$
   $y = \dfrac{3}{4}x$

   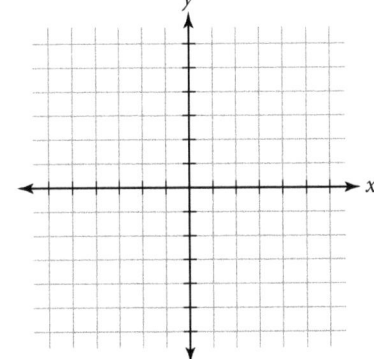

   b. $y = x^2 + 4x - 2$
   $y = 3x - 5$

   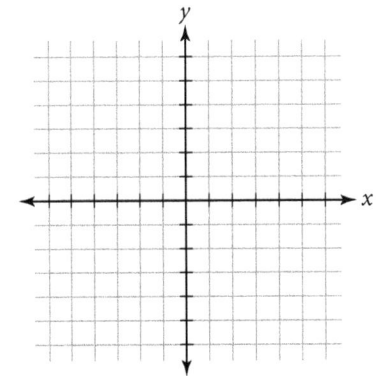

Systems of Inequalities

Name _____ Period _____ Date _____

1. Solve each inequality for y.

   a. $4x - 5y \leq 20$

   b. $3 + 2(x - 4y) > 12$

2. Graph each linear inequality on the coordinate plane.

   a. $y > \frac{1}{3}x - 3$

   b. $3x - 5y \geq 0$

   c. $4y - 2x \leq -8$

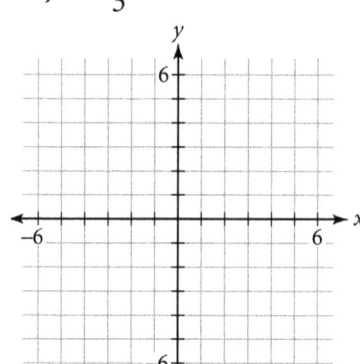 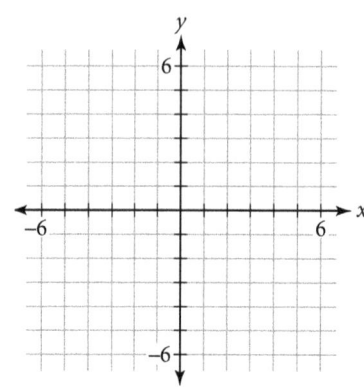 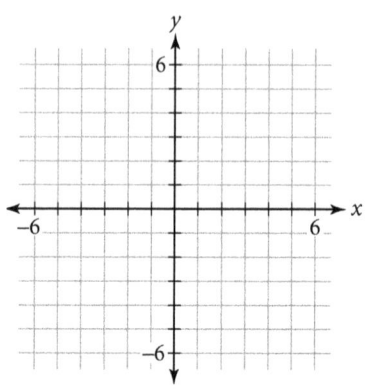

3. Graph the feasible region of each system of inequalities. Find the coordinates of each vertex.

   a. $\begin{cases} x - 3y \geq -11 \\ x - 3y \leq -2 \\ 1 \leq x \leq 4 \end{cases}$

   b. $\begin{cases} y \leq -|x| + 4 \\ y \leq 3x \\ x \geq 0 \\ y \geq 0 \end{cases}$

   c. $\begin{cases} y \leq \sqrt{25 - x^2} \\ y \geq 0.75x \\ x \geq 0 \end{cases}$

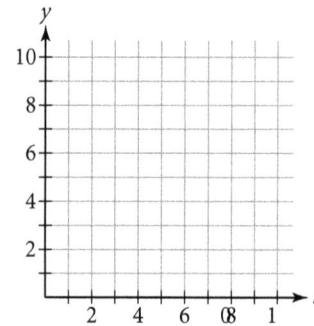 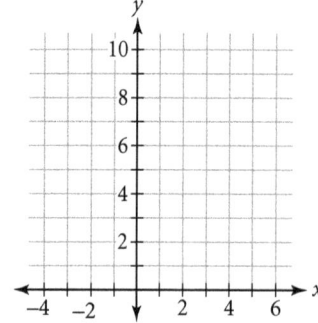 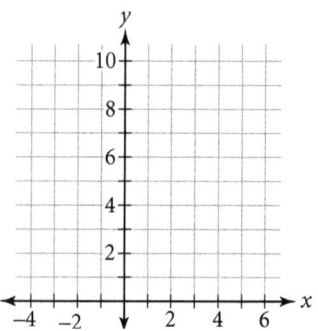

4. Leo is taking an algebra test containing computation problems worth 5 points each and application problems worth 8 points each. Leo needs to score at least 83 points on the test to maintain his B average. Let c represent the number of computation problems he answers correctly and a represent the number of application problems he answers correctly. Write an inequality to represent the constraint.

18

Systems of Equations and Inequalities
©2017 Kendall Hunt Publishing

Linear Programming

Name _____ Period _____ Date _____

1. A yoga teacher sells yoga mats for $25 each and yoga DVDs for $14 each. If x represents the number of mats she sells and y represents the number of DVDs she sells, write an expression that represents the amount of money she receives from selling mats and DVDs. Tell whether the expression should be minimized or maximized.

2. You would like to maximize profits at your bakery, which makes decorated sheet cakes for parties in two sizes, a full sheet and a half sheet. A batch of 12 full-sheet cakes takes 3.5 hours of oven time and 4 hours of decorating time, whereas a batch of 20 half-sheet cakes takes 5 hours of oven time and 2 hours of decorating time. The oven is available for a maximum of 21 hours a day, and the decorating room is available for 14 hours a day. Let x represent the number of batches of sheet cakes that the bakery produces in one day, and let y represent the number of batches of half-sheet cakes. The bakery makes a profit of $30 on each batch of full-sheet cakes and $35 on each batch of half-sheet cakes. The bakery must bake at least one batch of each kind to meet customer needs.

   a. Write a constraint about oven time.

   b. Write a constraint about decorating time.

   c. Write a system of inequalities that includes the constraints you have found and any commonsense constraints.

   d. Graph the feasible region and find the vertices.

   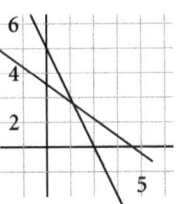

   e. Find the profit at each vertex.

   f. How many batches of each size of cake should the bakery make in one day to maximize profit? What is the maximum profit?

Larger Systems

Name _____ Period _____ Date _____

1. Solve each system using elimination.

   a. $x - 2y + z = -5$
   $x + y + z = 13$
   $-3x + y - 2z = -16$

   b. $-2x + 2y - z + 5 = 0$
   $2x - y + 3z + 6 = 0$
   $x + 2z + 5 = 0$

2. Solve each system using substitution.

   a. $3x - y - z = 120$
   $y - 2z = 30$
   $x + y + z = 180$

   b. $x - 6y - 2z = -8$
   $-x + 5y + 3z = 2$
   $3x - 2y - 4z = 18$

3. Determine if the given point is a solution for the system.

   a. Is the point $\left(2, \frac{1}{2}, -3\right)$ a solution for the system:
   $x + 2y + z = 0$
   $3x - 4y + 5z = -11$
   $-2x - 8y - 3z = 1$

   b. Is the point $(5, 6, 3)$ a solution for the system:
   $2x + y - 2z = 1$
   $6x + 2y - 4z = 3$
   $4x - y + 3z = 5$

Functions and Relations

Interpreting Graphs

Name _____ Period _____ Date _____

1. Describe the pattern of the graph of each of the following situations as the graphs are read from left to right as increasing, decreasing, increasing and then decreasing, or decreasing and then increasing.

   a. The height of a child at birth and on each birthday from age 1 to age 6

   b. The height of a ball that is thrown upward from the top of a building from the time it is thrown until it hits the ground

2. For each of the situations described in Exercise 1, describe the real-world meaning of the vertical intercept of the graph.

3. Sketch a graph to match the description below.
   Increasing rapidly at a constant rate, then suddenly becoming constant, then decreasing rapidly at a constant rate

4. Sketch what you think is a reasonable graph for each relationship described. In each situation, identify the variables and label your axes appropriately.

   a. The temperature of a hot drink sitting on your desk

   b. Your speed as you cycle up a hill and down the other side

Function Notation

Name _____ Period _____ Date _____

1. Determine whether or not each graph represents a function. Explain how you know.

   a.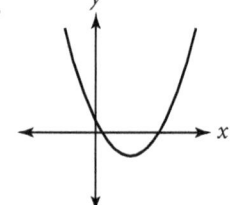

   b.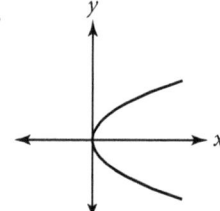

   c.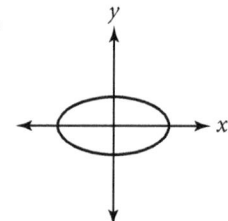

2. Find each of the indicated function values.

   a. If $f(x) = -\sqrt{4x+1}$, find $f\left(-\frac{1}{4}\right)$, $f(0)$, $f(0.75)$, $f(2)$, and $f(12)$.

   b. If $f(x) = \frac{2}{x-4}$, find $f(-4)$, $f(0)$, $f(5)$, $f(8)$, and $f(24)$.

3. Use the graph at right to find each of the following.

   a. $f(3) + f(-3)$

   b. $f(f(10))$

   c. x when $f(x) = -3$

   d. x when $f(x-3) = 35$

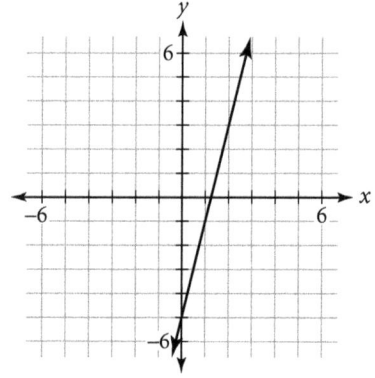

4. Define variables and write a function that describes each situation.

   a. You drive on an interstate highway with your cruise control set at 65 miles per hour and do not need to stop or alter your speed.

   b. You rent a small moving van to move your belongings to your new apartment. The rental company charges $45 a day plus $0.22 a mile to rent the van.

Lines in Motion

Name _____ Period _____ Date _____

1. Describe how each graph translates the graph of $y = f(x)$.

    a. $y = f(x) - 3$ b. $y = f(x + 6)$ c. $y = 5 + f(x - 7)$

2. Find each of the following.

    a. $f(x - 2)$ if $f(x) = -4x$ b. $3 + f(x + 4)$ if $f(x) = 2x$

    c. $f(x - 5)$ if $f(x) = 2x + 1$ d. $3 + f(x + 6)$ if $f(x) = 8 - x$

3. Write an equation for each line.

    a. The line $y = -1.2x$ translated right 3 units

    b. The line $y = -x$ translated up 5 units and left 2 units

    c. The line $y = \frac{1}{2}x$ translated down 1 unit and right 4 units

4. The graph of $y = f(x)$ is shown at right. Write an equation for each related graph showing how the function has been translated.

a.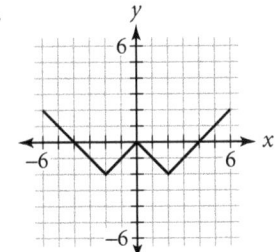

b.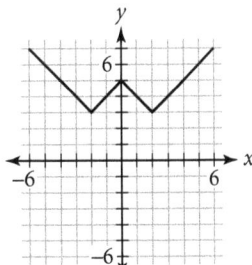

c.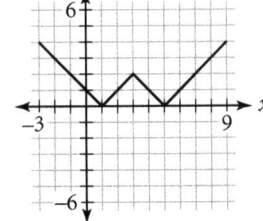

d.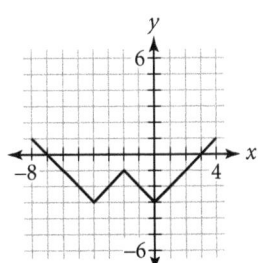

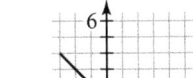

Translations and the Quadratic Family

Name _____ Period _____ Date _____

1. Describe the translations of the graph of $y = x^2$ needed to produce the graph of each equation.

   a. $y = x^2 - 6$

   b. $y = (x + 5)^2$

   c. $y = (x - 3)^2 - 9$

2. Find the vertex of each parabola.

   a. $y = x^2 + 3$

   b. $y = (x - 2)^2$

   c. $y = -8 + (x + 5)^2$

3. Each parabola described is the graph of $y = x^2$. Write an equation for each parabola and sketch its graph.

   a. The parabola is translated horizontally -3 units.

   b. The parabola is translated vertically 1 unit.

   c. The parabola is translated horizontally 2 units and vertically -3 units.

4. Describe what happens to the graph of $y = x^2$ in the following situations.

   a. y is replaced with $(y + 1)$.

   b. x is replaced with $(x - 5)$.

5. Solve.

   a. $x^2 + 6 = 31$

   b. $x^2 - 12 = 52$

   c. $(x - 3)^2 = 100$

   d. $(x + 7)^2 = 144$

   e. $(x + 4)^2 - 5 = 31$

   f. $-20 + (x - 5)^2 = 3$

Reflections and the Square Root Family

Name _____ Period _____ Date _____

1. Describe what happens to the graph of $y = \sqrt{x}$ in each of the following situations.

   a. x is replaced with $(x + 6)$.

   b. y is replaced with $(y - 5)$.

   c. y is replaced with $(y + 1)$.

   d. x is replaced with $(x - 8)$.

2. Each graph below is a transformation of the graph of either the parent function $y = x^2$ or the parent function $y = \sqrt{x}$. Write an equation for each graph.

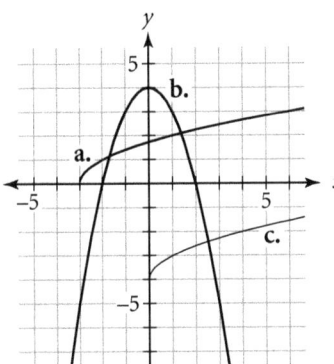
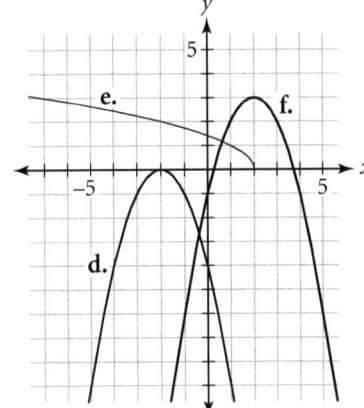

3. Given the graph of $y = f(x)$ at right, draw a graph of each of these related functions.

   a. $y = -f(x)$ b. $y = f(-x)$ c. $y = -f(-x)$

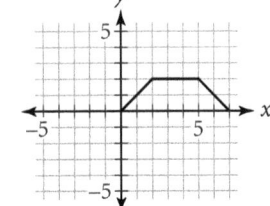

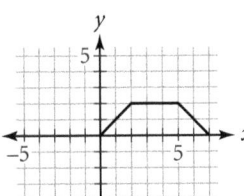
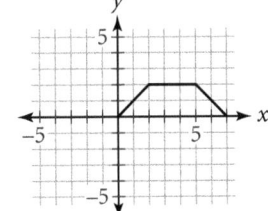
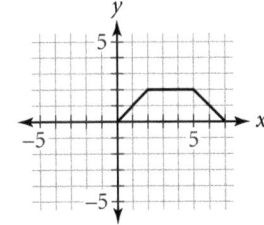

Dilations and the Absolute-Value Family

Name _____ Period _____ Date _____

1. Each graph is a transformation of one of the parent functions you've studied. Write an equation for each graph.

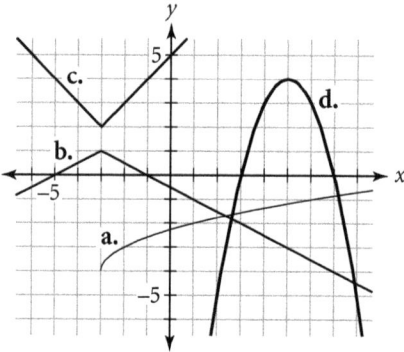

2. Describe the transformations of the graph of $y = |x|$ needed to produce the graph of each equation.

   a. $y = |x - 3|$

   b. $\dfrac{y}{1.5} = \dfrac{x}{2}$

   c. $\dfrac{y + 4}{2} = |x - 1|$

3. Find the vertex of the graph of each equation in Exercise 2 and sketch the graph.

4. Solve.

   a. $|x - 5| - 7 = 0$

   b. $3|x - 5| - 2 = 10$

   c. $\left|\dfrac{x}{2}\right| + 5 = 12$

5. Solve each equation for y.

   a. $\dfrac{y}{2} = \left|\dfrac{x}{4}\right|$

   b. $\dfrac{y - 3}{2} = (x + 1)^2$

   c. $\dfrac{y + 1}{-3} = \sqrt{x + 2}$

Transformations and the Circle Family

Name _____ Period _____ Date _____

1. Write an equation for each circle.

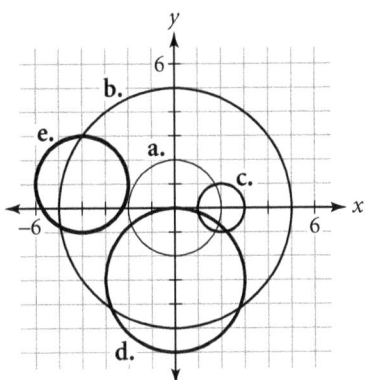

2. If $f(x) = \sqrt{1 - x^2}$, write an equation for each of the following related functions.

   a. $-f(x)$ b. $f(-x)$ c. $2f(x)$ d. $f(2x)$

3. Without graphing, find the x- and y-intercepts of the graph of each equation.

   a. $y = -\sqrt{1 - x^2}$ b. $y = -2\sqrt{1 - x^2}$ c. $y = \sqrt{1 - (2x)^2}$

   d. $y = -2\sqrt{1 - (4x)^2}$ e. $y = -\sqrt{1 - \left(\frac{x}{3}\right)^2}$ f. $y = 2\sqrt{1 - \left(\frac{x}{4}\right)^2}$

4. Write an equation for each transformation of the unit circle, and identify its graph as a circle or an ellipse. Then sketch the graph.

   a. Replace x with $\frac{x}{2}$ and y with $\frac{y}{2}$. b. Replace x with $\frac{x}{4}$ and y with $\frac{y}{3}$.

Exponential, Power, and Logarithmic Functions

Exponential Functions

Name _____ Period _____ Date _____

1. Evaluate each function at the given value. Round to four decimal places if necessary.

   a. $r(t) = 325(1 + 0.035)^t$, $t = 8$

   b. $j(x) = 59.5(1 - 0.095)^x$, $x = 10$

2. Record the next three terms for each sequence. Then write an explicit function for the sequence.

   a. $a_0 = 12$
      $a_n = 0.8a_{n-1}$ where $n \geq 1$

   b. $u_0 = 50.5$
      $u_n = 2.1u_{n-1}$ where $n \geq 1$

3. Evaluate each function at $x = 0$, $x = 1$, and $x = 2$ and write a recursive formula for the pattern. Then, indicate whether each equation is a model for exponential growth or decay. Describe the end behavior of each function.

   a. $f(x) = 2000(0.9)^x$

   b. $f(x) = 3000(1 + 0.001)^x$

   c. $f(x) = 0.1(1 - 0.5)^x$

4. Calculate the ratio of the second term to the first term, and express the answer as a decimal value. State the percent increase or decrease.

   a. 80, 60

   b. 36, 32

   c. 63, 100.8

5. Rohit bought a new car for $17,500. The value of the car is depreciating at a rate of 16% a year.

   a. Write a recursive formula that models this situation. Let u_0 represent the purchase price, u_1 represent the value of the car after 1 year, and so on.

   b. Define variables and write an exponential equation that models this situation.

Properties of Exponents and Power Functions

Name _____ Period _____ Date _____

1. Rewrite each expression as a fraction without exponents or as an integer. Using your calculator, verify that your answer is equivalent to the original expression.

    a. 3^{-2}

    b. 7^{-3}

    c. -4^{-4}

    d. $(-5)^{-3}$

    e. $-\left(\dfrac{3}{5}\right)^{-2}$

    f. $\left(-\dfrac{5}{6}\right)^{-2}$

2. Rewrite each expression in the form x^n or ax^n.

    a. $4x^0 \cdot 9x^8$

    b. $(8x^{-6})(-15x^{-14})$

    c. $\dfrac{x^9}{x^{-9}}$

    d. $\dfrac{-88x^{10}}{-8x^3}$

    e. $\left(\dfrac{-35x^7}{-7x^2}\right)^3$

    f. $\left(\dfrac{40x^{-8}}{-8x^{-2}}\right)^{-3}$

3. Solve.

    a. $2^x = \dfrac{1}{32}$

    b. $125^x = 25$

    c. $\left(\dfrac{4}{9}\right)^x = \dfrac{81}{16}$

4. Solve each equation for positive values of x. If answers are not exact, approximate to two decimal places.

    a. $6x^{1.5} = 80$

    b. $20x^{1/2} - 8 = 4.5$

    c. $5x^{-1/3} = 0.06$

    d. $8x^9 = 6x^6$

    e. $15x^{-3} = 10x^{-2}$

    f. $200x^{-1} = 125x^{-3}$

Rational Exponents and Roots

Name _____ Period _____ Date _____

1. Identify each function as a power function, an exponential function, or neither of these. (The function may be translated, stretched, or reflected.)

   a. $f(x) = 0.5x^3 - 4$
   b. $f(x) = \frac{1}{3^x}$
   c. $f(x) = \frac{1}{x} + 2$

2. Rewrite each expression in the form b^x in which x is a rational exponent.

   a. $\sqrt{c^3}$
   b. $(\sqrt[3]{d})^4$
   c. $\frac{1}{\sqrt{r^5}}$

3. Solve each equation for positive values of x. If answers are not exact, approximate to the nearest hundredth.

   a. $\sqrt[5]{x^3} = 27$
   b. $\frac{1}{\sqrt{x}} = 0.77$
   c. $4\sqrt[3]{x} + 18 = 32$

   d. $\sqrt[5]{x^3} - 23 = -15$
   e. $\sqrt[3]{4x^2} + 8.5 = 19.8$
   f. $\sqrt[8]{x^5} = 12.75$

4. Each of the following graphs is a transformation of the power function $y = x^{3/2}$. Write the equation for each curve.

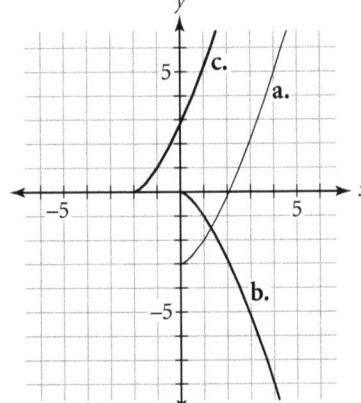
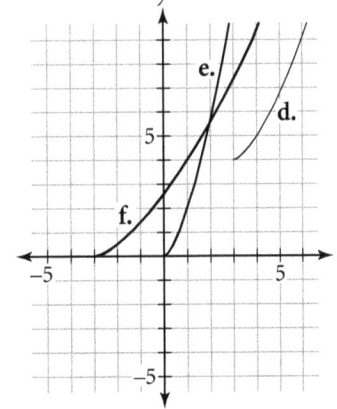

Applications of Exponential and Power Equations

Name _____ Period _____ Date _____

1. Solve each equation for positive values of x. If answers are not exact, approximate to the nearest hundredth.

   a. $\sqrt[3]{x} = 2.6$

   b. $x^{-1/4} = 0.2$

   c. $0.75x^5 - 8 = -3$

   d. $4(x^{5/6} + 7) = 159$

   e. $224 = 200\left(1 + \frac{x}{4}\right)^9$

   f. $1500\left(1 + \frac{x}{12}\right)^{6.5} = 1525$

2. Rewrite each expression in the form ax^n.

   a. $(8x^9)^{2/3}$

   b. $(81x^{12})^{3/4}$

   c. $(49x^{-10})^{1/2}$

   d. $(-27x^{-9})^{4/3}$

   e. $(100{,}000x^{10})^{3/5}$

   f. $(-125x^{-15})^{1/3}$

3. Give the average annual rate of inflation for each situation described. Give your answers to the nearest tenth of a percent.

   a. The cost of a movie ticket increased from $6.00 to $8.50 over 10 years.

   b. The monthly rent for Hector's apartment increased from $650 to $757 over 4 years.

4. The population of a small town has been declining because jobs have been leaving the area. The population was 23,000 in 2002 and 18,750 in 2007. Assume that the population is decreasing exponentially.

   a. Define variables and write an equation that models the population in this town in a particular year.

   b. Use your model to predict the population in 2010.

   c. According to your model, in what year will the population first fall below 12,000?

Building Inverses of Functions

Name _____ Period _____ Date _____

1. Each of the functions below has an inverse that is also a function. Find four points on the graph of each function f, using the given values of x. Use these points to find four points on the graph of f^{-1}.

   a. $f(x) = 3x - 4;\ x = -2,\ 0,\ \frac{4}{3},\ 4$

   b. $f(x) = x^3 - 2;\ x = -3,\ -1,\ 2,\ 5$

2. For each function below, determine whether or not the inverse of the function is a function. Find the equation of the inverse and graph both equations on the same axes.

   a. $y = -2x + 5$

   b. $y = |x|$

   c. $y = x^2 - 4$

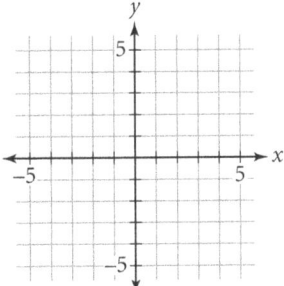
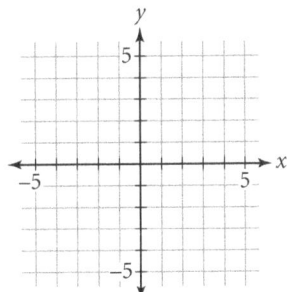
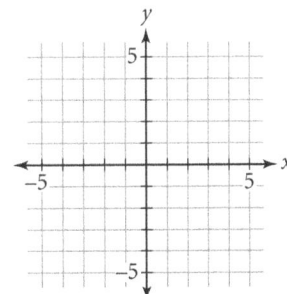

3. Balloons and Laughs Inc. is a small company that entertains at children's birthday parties. B & L uses a complicated formula to calculate its prices, taking into account all of its costs. The price equation is $p(x) = 4\sqrt[3]{(8x + 3)^2} + 25$, where x is the number of person-hours supplied for the party at a price of $p(x)$. For example, if $x = 4$, four clowns will come for one hour, two clowns will come for two hours, or one clown will come for four hours.

   a. What is the price if two clowns come to a party for 90 minutes?

   b. Many customers want to know what they can get for a particular amount of money. Rewrite the price equation for B & L so that the company can input the amount of money a customer wants to spend and the output will be the number of person-hours he or she will get for the money. Call the new function $p^{-1}(x)$.

   c. B & L's Ultimate Party costs $125. How many person-hours do you get at an Ultimate Party?

Logarithms

Name _____ Period _____ Date _____

1. Rewrite each logarithmic equation in exponential form using the definition of logarithm. Then solve for x.

   a. $\log_3 \frac{1}{81} = x$

   b. $\log_x \sqrt[4]{12} = \frac{1}{4}$

   c. $x = \log_4 32$

   d. $\log x = 1$

   e. $3 = \log_x 125$

   f. $\ln x = 5$

2. Find the exact value of each logarithm without using a calculator. Write answers as integers or fractions in lowest terms.

   a. $\log_3 81$

   b. $\log_5 \sqrt{5}$

   c. $\log_3 \frac{1}{3}$

   d. $\log_2 \frac{1}{32}$

   e. $\log_8 4$

   f. $\log 1{,}000{,}000{,}000$

3. Each graph is a transformation of either $y = 10^x$ or $y = \log x$. Write the equation for each graph.

   a.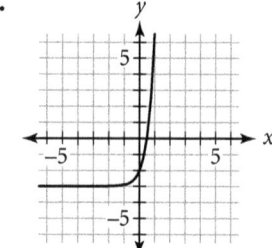

   b.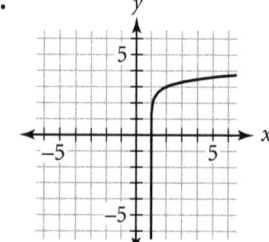

   c.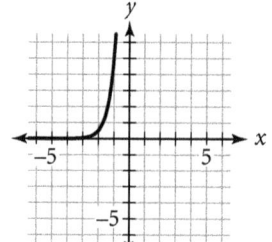

4. Use technology to solve each equation. (Round to four decimal places.)

   a. $\log_5 120 = x$

   b. $\log_3 0.9 = x$

   c. $4^x = 99$

   d. $6^x = 729$

   e. $7^x = 4.88$

   f. $e^x = 15$

Arithmetic Series

Name _____   Period _____   Date _____

1. List the first six terms of each arithmetic sequence and identify the common difference.

   a. $u_1 = 5$
      $u_n = u_{n-1} + 6$ where $n \geq 2$

   b. $a_1 = 7.8$
      $a_n = a_{n-1} - 2.3$ where $n \geq 2$

2. Write each expression as a sum of terms and calculate the sum.

   a. $\sum_{n=1}^{3}(n-5)$

   b. $\sum_{n=1}^{4}(3n-7)$

   c. $\sum_{n=1}^{5}(2n^2+5)$

3. Find the indicated values.

   a. u_{18} if $u_n = \frac{2}{3}n + \frac{3}{4}$

   b. S_{10} if $u_n = 3n - 6$

   c. $\sum_{n=51}^{100}(9n-81)$

4. There are 22 rows of seats in a high school auditorium. There are 17 seats in the front row, and each of the other row has two more seats than the row directly in front of it.

   a. List the first six terms of the sequence that describes the number of seats in each row, starting with the front row.

   b. Write a recursive formula for this sequence.

   c. Write an explicit formula for the number of seats in row n.

   d. The rows are identified with letters to help people who attend performances in the auditorium find their seats. If the front row is row A, how many seats are there in row M?

   e. How many seats are there in the back row?

   f. How many seats are there in the auditorium?

Discovering Advanced Algebra More Practice Your Skills

Partial Sums of Geometric Series

Name _____ Period _____ Date _____

1. For each partial sum equation, identify the first term, the common ratio, and the number of terms.

   a. $\dfrac{15}{1-0.6} - \dfrac{15}{1-0.6} \cdot 0.6^5 = 34.584$

   b. $\dfrac{30 - 0.1171875}{1 - 0.5} = 59.765625$

2. Consider the geometric sequence 187.5, 75, 30, 12, ...

   a. What is the tenth term?

   b. Which term is the first one smaller than 1?

   c. Find u_9.

   d. Find S_9.

3. Find the first term and the common ratio or common difference of each series. Then find the partial sum.

   a. $5 + 6.2 + 7.4 + \cdots + 17$

   b. $150 - 30 + 6 - 1.2 + \cdots + 0.000384$

   c. $\sum\limits_{n=1}^{15} 12.5(1.1)^{n-1}$

   d. $\sum\limits_{n=1}^{50} (72 - 3.5n)$

4. Find the missing values.

   a. $u_1 = 4$, $r = 3$, $S_{10} = $ _____

   b. $u_1 = 2$, $r = 0.8$, $S\text{\_\_\_\_\_} = 6.7232$

   c. $u_1 = $ _____, $r = 1.1$, $S_6 = 92.58732$

   d. $u_1 = 10$, $r = $ _____, $S_{12} = 19.99511719$

5. Suppose that you rent an apartment for $750 a month. Each year, your landlord raises the rent by 5%.

   a. If you make a list of your monthly rent over several years, does this form an arithmetic or geometric sequence? If the sequence is arithmetic, give the common difference; if it is geometric, give the common ratio.

   b. To the nearest dollar, what is your monthly rent during the fifth year you rent the apartment?

   c. To the nearest ten dollars, what is the total amount of rent that you paid during the first five years you rented the apartment?

Quadratic Functions and Relations

Equivalent Quadratic Forms

Name _____ Period _____ Date _____

1. Identify each quadratic function as being in general form, vertex form, factored form, or none of these forms.

   a. $y = 3x^2 - 4x + 5$
   b. $y = (x - 2.5)^2 + 7.5$
   c. $y = -1.5x(x - 2)$

2. Convert each quadratic function to general form.

   a. $y = (x - 3)^2$
   b. $y = -5(x + 3)(x - 2) - 30$
   c. $y = 3(x - 1.5)^2 - 10$

3. Find the vertex of the graph of each quadratic function.

   a. $y = -x^2$
   b. $y = -(x - 1)^2 + 6$
   c. $y = 6.5 + 0.5(x + 4)^2$

4. Find the zeros of each quadratic function.

   a. $y = -2(x - 1)(x + 6)$
   b. $y = 0.5x(x - 5)$
   c. $y = (x - 7.5)^2$

5. Consider this table of values generated by a quadratic function.

x	-3	-2.5	-2	-1.5	-1	-0.5	0
y	-0.5	-3	-4.5	-5	-4.5	-3	-0.5

   a. What is the line of symmetry for the graph of the quadratic function?

   b. Identify the vertex of the graph of this quadratic function, and determine whether it is a maximum or a minimum.

   c. Use the table of values to write the quadratic function in vertex form.

Completing the Square

Name _____ Period _____ Date _____

1. Factor each quadratic expression.

    a. $x^2 + 10x + 25$ b. $x^2 - x + \frac{1}{4}$ c. $9x^2 - 24xy + 16y^2$

2. What value is required to complete the square?

    a. $x^2 - 18x +$ _____   b. $x^2 - 5x +$ _____   c. $x^2 + 4.3x +$ _____

3. Convert each quadratic function to vertex form by completing the square.

    a. $y = x^2 + 14x + 50$ b. $y = 5x^2 - 10x - 3$ c. $y = 2x^2 + 5x$

4. Find the vertex of the graph of each quadratic function, and state whether the vertex is a maximum or a minimum.

    a. $y = (x - 2)(x + 6)$ b. $y = -3.5x^2 - 7x$ c. $y = x^2 + 9x - 10$

5. Rewrite each expression in the form $ax^2 + bx + c$, and then identify the coefficients a, b, and c.

    a. $-6 + 3x^2 + 6x + 8$ b. $-2x(x - 8)$ c. $(2x - 3)(x + 5)$

6. A ball is thrown up and off the roof of a 75 m tall building with an initial velocity of 14.7 m/s.

    a. Let t represent the time in seconds and h represent the height of the ball in meters. Write an equation that models the height of the ball.

    b. At what time does the ball reach maximum height? What is the ball's maximum height?

    c. At what time does the ball hit the ground?

The Quadratic Formula

Name _____ Period _____ Date _____

1. Evaluate each expression. Round your answers to the nearest thousandth.

   a. $\dfrac{-6 + \sqrt{6^2 - 4(1)(-5)}}{2(1)}$

   b. $\dfrac{4 - \sqrt{(-4)^2 - 4(2)(1)}}{2(2)}$

   c. $\dfrac{5 + \sqrt{(-5)^2 - 4(4)(-3)}}{2(4)}$

   d. $\dfrac{-10 - \sqrt{10^2 - 4(2)(5)}}{2(2)}$

2. Solve by any method. Give your answers in exact form.

   a. $x^2 + 3x - 10 = 0$
   b. $2x^2 - 5x = 12$
   c. $25x^2 - 49 = 0$

   d. $4x^2 + 7x - 1 = 0$
   e. $x^2 = 5.8x$
   f. $x^2 - 48 = 0$

3. Use the quadratic formula to find the zeros of each function. Then, write each equation in factored form, $y = a(x - r_1)(x - r_2)$, where r_1 and r_2 are the zeros of the function.

   a. $y = x^2 + 5x - 24$
   b. $y = 2x^2 - 8x + 6$
   c. $y = 4x^2 + 2x - 2$

4. Write a quadratic function in general form that satisfies the given conditions.

   a. $a = -1$; x-intercepts of graph are -4 and -2

   b. x-intercepts of graph are 0 and 13; graph contains point (2, 22)

   c. x-intercept of graph is 4.8; y-intercept is -5.76

Discovering Advanced Algebra More Practice Your Skills
©2017 Kendall Hunt Publishing

Complex Numbers

Name _____ Period _____ Date _____

1. Add, subtract, or multiply.

   a. $(-5 + 6i) - (1 - i)$

   b. $(-2.4 - 5.6i) + (5.9 + 1.8i)$

   c. $-4i(-6 + i)$

   d. $(2.5 + 1.5i)(3.4 - 0.6i)$

2. Find the conjugate of each complex number.

   a. $5 - 4i$

   b. $7i$

   c. $-3.25 + 4.82i$

3. Use the definitions of $i = \sqrt{-1}$ and $i^2 = -1$ to rewrite each power of i as 1, i, -1, or $-i$.

   a. i^7

   b. i^{10}

   c. i^{16}

4. Solve each equation. Use substitution to check your solutions. Label each solution as real, imaginary, and/or complex.

   a. $x^2 - 2x + 5 = 0$

   b. $x^2 + 7 = 0$

   c. $x(x - 5) = 1$

   d. $x^2 + x + 1 = 0$

   e. $4x^2 + 9 = 0$

   f. $(x + 7)(x - 3) = 5 - 2x$

5. Write a quadratic function in general form that has the given zeros and leading coefficient of 1.

   a. $x = -4$, $x = 7$

   b. $x = 11i$, $x = -11i$

   c. $x = -2 + 3i$, $x = -2 - 3i$

Solving Quadratic Equations

Name _____ Period _____ Date _____

1. Write each equation in standard form $ax^2 + bx + c = 0$. What are the values of a, b, and c in each equation?

   a. $5x^2 + 7x = 5$
   b. $6 - x^2 = 8x$
   c. $3x^2 = 48$

2. For each quadratic equation, find the value of the discriminant and classify what type of solutions the equation will have. You do not need to find the solutions. Use a graph to support your answer.

   a. $x^2 - 5x + 2 = 0$
   b. $0.5x^2 - 2x + 2 = 0$
   c. $3x^2 - x + 1 = 0$

3. For each of the following equations, tell which method is the most appropriate method to use to solve the equation and why.

   a. $x^2 - 11x + 24 = 0$
   b. $2x^2 - 4x - 9 = 0$
   c. $-5x^2 + 11x - 1631 = 0$

4. Solve each equation by factoring.

   a. $x^2 + 4x - 21 = 0$
   b. $9x^2 - 15x + 4 = 0$
   c. $x^2 - 121 = 0$

5. Solve each equation by completing the square.

   a. $x^2 + 4x - 21 = 0$
   b. $x^2 + 8x + 15 = 0$
   c. $x^2 + 3x - 28 = 0$

6. Solve each equation using the quadratic formula.

   a. $3x^2 - 7x + 2 = 0$
   b. $5x^2 - 8x + 3 = 0$
   c. $3x^2 + 2x - 7 = 0$

7. Solve the equation $(x + 3)^2 - 2(x + 3) + 1 = 0$.

Discovering Advanced Algebra More Practice Your Skills
©2017 Kendall Hunt Publishing

Solving Radical Equations

Name _____ Period _____ Date _____

1. Square each expression.

   a. $\sqrt{2x-1}$

   b. $\sqrt{3x-4}$

   c. $2\sqrt{x-5}$

   d. $\sqrt{12x^2} + \sqrt{3x^2}$

2. Determine the number of solutions for each equation by graphing.

   a. $\sqrt{5x^2+5} = 5$

   b. $\sqrt{x+2} = -x$

   c. $\sqrt{2x+6} = \sqrt{2x-5}$

   d. $3 + \sqrt{x-1} = 5$

3. Solve each equation.

   a. $\sqrt{x} + 7 = 0$

   b. $8 - 4\sqrt{5x} = 0$

   c. $2\sqrt{x-2} = \sqrt{7-x}$

   d. $4 + \sqrt{x-3} = 11$

4. Solve each equation.

   a. $x - 1 = \sqrt{x+5}$

   b. $\sqrt{3x+1} = 1 + \sqrt{x+4}$

Using the Distance Formula

Name _____ Period _____ Date _____

1. Find the exact distance between each pair of points.

   a. (0, 0) and (5, 12)

   b. (2, 8) and (6, 11)

   c. (−2, 5) and (2, 7)

   d. (4, −7) and (8, −15)

   e. (3a, 8) and (−2a, 5)

   f. $\left(\frac{1}{2}, \frac{1}{4}\right)$ and $\left(-\frac{1}{2}, \frac{9}{4}\right)$

2. Make a sketch of the situation and find the possible values of x or y.

   a. The distance between the points (−4, 6) and (2, y) is 10 units.

   b. The distance between the points (4, −5) and (x, 3) is 11 units.

3. Make a sketch of the situation and find an equation of the locus of points that satisfies the given condition.

   a. The points that are 5 units from (−2, 3)

Discovering Advanced Algebra More Practice Your Skills
©2017 Kendall Hunt Publishing

b. The points that are equidistant from (0, 0) and (2, 5)

c. The points that are twice as far from (−9, 0) as they are from (0, 0)

4. Find the center and radius of each circle.
   a. $x^2 + y^2 + 6x + 5 = 0$ b. $2x^2 + 2y^2 - 8x + 12y - 46 = 0$

Parabolas

Name _____ Period _____ Date _____

1. For each parabola described, use the information given to find the location of the missing feature. It may help to draw a sketch.

   a. If the vertex is (0, 0) and the focus is (4, 0), where is the directrix?

   b. If the vertex is (5, 0) and the directrix is $x = 1.5$, where is the focus?

   c. If the focus is (2, −3) and the directrix is $x = -1$, where is the vertex?

2. Find the vertex of each parabola and state whether the parabola opens upward, downward, to the right, or to the left. Also give the equation of the axis of symmetry.

   a. $y = x^2 - 5$　　　b. $y = -4x^2$　　　c. $x = 2y^2 + 1$

   d. $x = -(y - 3)^2$　　　e. $y + 2 = -(x + 1)^2$　　　f. $\left(\dfrac{y-4}{2}\right)^2 = \dfrac{x+5}{4}$

3. Write an equation in standard form for each parabola.

   a. 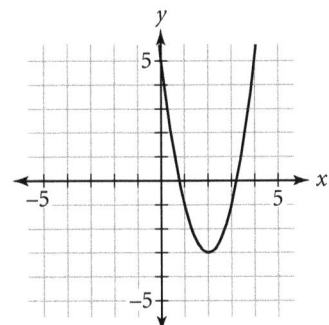　　b. 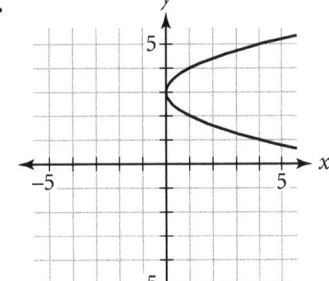　　c.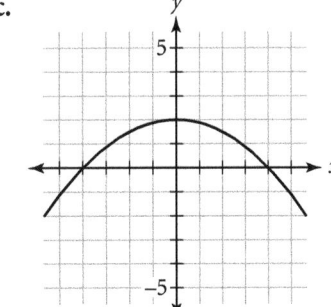

Polynomial and Rational Functions

Polynomials

Name _____ Period _____ Date _____

1. Identify the degree of each polynomial.

   a. $x^5 - 1$
   b. $0.2x - 1.5x^2 + 3.2x^3$
   c. $250 - 16x^2 + 20x$

2. Determine which of the expressions are polynomials. For each polynomial, state its degree and write it in general form. If it is not a polynomial, explain why not.

   a. $0.2x^3 + 0.5x^4 + 0.6x^2$
   b. $x - \dfrac{1}{x^2}$
   c. 25

3. The figures below show why the numbers in the sequence 1, 3, 6, 10, . . . are called *triangular numbers*.

    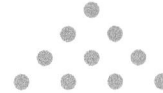

   a. Complete the table.

n	1	2	3	4	5	6	7
*n*th triangular number	1	3	6	10			

   b. Calculate the finite differences for the completed table.

   c. What is the degree of the polynomial function that you would use to model this data set?

   d. Write a polynomial function *t* that gives the *n*th triangular number as a function of *n*. (Hint: Create and solve a system of equations to find the coefficients.)

Discovering Advanced Algebra More Practice Your Skills

Factoring Polynomials

Name _____ Period _____ Date _____

1. Without graphing, find the x-intercepts and the y-intercept for the graph of each equation.

   a. $y = -(x-8)^2$

   b. $y = 3(x+4)(x+2)$

   c. $y = 0.75x(x-2)(x+6)$

2. Write the factored form of the quadratic function for each graph. Don't forget the vertical scale factor.

   a.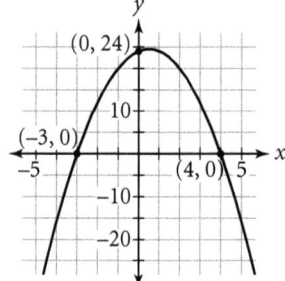

   b.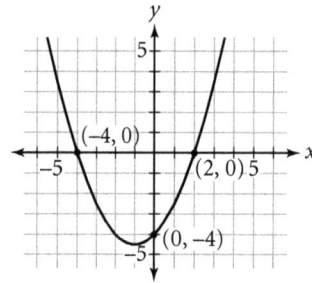

3. Convert each polynomial function to general form.

   a. $y = -2(x-2.5)(x+2.5)$

   b. $y = -0.5(x+3)^2$

   c. $y = -x(x+12)(x-12)$

4. Write each polynomial as a product of factors. Some factors may include irrational numbers.

   a. $x^2 - 14x + 49$

   b. $x^3 - 3x^2 + 2x$

   c. $x^2 + 169$

   d. $x^2 - 15$

   e. $x^4 - 10x^2 + 9$

   f. $3x^3 + 3x^2 - 30x + 24$

5. Sketch a graph for each situation if possible.

   a. A quadratic function with two real zeros, whose graph has the line $x = 2$ as its axis of symmetry

   b. A cubic function with three real zeros, whose graph has a positive y-intercept

Higher-Degree Polynomials

Name _____ Period _____ Date _____

1. Refer to these two graphs of polynomial functions.

   i. 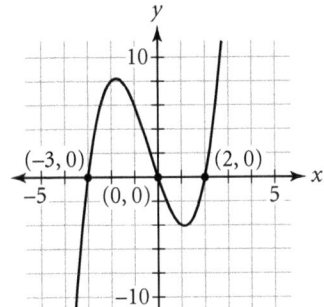   ii.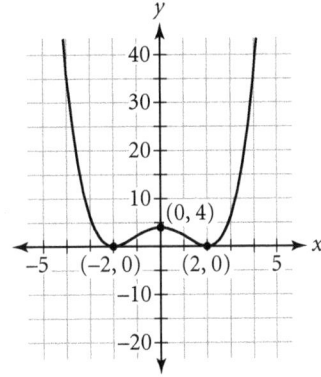

   a. Identify the zeros of each function.

   b. Give the coordinates of the y-intercept of each graph.

   c. Identify the lowest possible degree of each polynomial function.

   d. Write the factored form for each polynomial function. Check your work by graphing on your calculator.

2. Write a polynomial function with the given features.

   a. A quadratic function whose graph has vertex $(3, -8)$, which is a minimum, and two x-intercepts, one of which is 5

   b. A fourth-degree polynomial function with two double roots, 0 and 2, and whose graph contains the point $(1, -1)$

3. Write the lowest-degree polynomial function that has the given set of zeros and whose graph has the given y-intercept. Write each polynomial function in factored form. Give the degree of each function.

   a. Zeros: $x = -3$, $x = 5$; y-intercept: -30

   b. Zeros: $x = \pm 2i$, $x = -2$ (double root), $x = 5$; y-intercept: 80

More About Finding Solutions

Name _____ Period _____ Date _____

1. Divide.

   a. $x - 2 \overline{) 3x^3 - 8x^2 - 11x + 30}$

   b. $x - 4 \overline{) x^4 - 13x^2 - 48}$

2. Varsha started out dividing two polynomials by synthetic division this way:

   $\underline{-3 |} -3 \quad -5 \quad 0 \quad -35 \quad 7$

   a. Identify the dividend and divisor.

   b. Write the numbers that will appear in the second line of the synthetic division.

   c. Write the numbers that will appear in the last line of the synthetic division.

   d. Write the quotient and remainder for this division.

3. In each division problem, use the polynomial that defines P as the dividend and the binomial that defines D as the divisor. Write the result of the division in the form $P(x) = D(x) \cdot Q(x) + R$, where the polynomial that defines Q is the quotient and R is an integer remainder. (It is not necessary to write the remainder if $R = 0$.)

   a. $P(x) = 2x^2 - 9x + 2; D(x) = x - 5$

   b. $P(x) = 2x^3 - 5x^2 + 8x - 5; D(x) = x - 1$

4. Make a list of the possible rational roots of each equation.

   a. $x^3 + x^2 - 10x + 8 = 0$

   b. $2x^3 - 3x^2 - 17x + 30 = 0$

5. Find all the zeros of each polynomial function. Then write the function in factored form.

   a. $y = x^3 - 5x^2 + 9x - 45$

   b. $y = 6x^3 + 17x^2 + 6x - 8$

Introduction to Rational Functions

Name _____ Period _____ Date _____

1. Write an equation and graph each transformation of the parent function $f(x) = \frac{1}{x}$.

   a. Translate the graph left 3 units and down 4 units.

   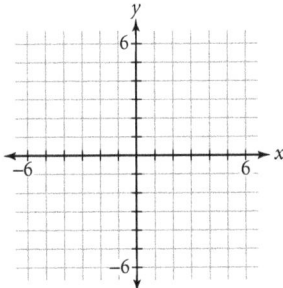

   b. Vertically dilate the graph by a scale factor of 3.

   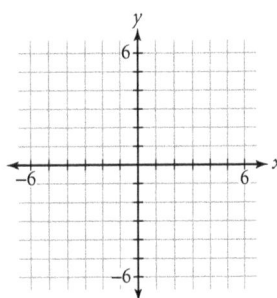

2. Write equations for the asymptotes of each hyperbola.

   a. $y = \frac{2}{x}$

   b. $y = \frac{1}{x+3}$

   c. $y = -\frac{3}{x}$

   d. $y = \frac{1}{x} + 5$

   e. $y = \frac{1}{x-2} - 6$

   f. $y = \frac{4}{x+2} - 1$

3. Solve.

   a. $\frac{6}{x-5} = -2$

   b. $\frac{4}{2x+5} = -\frac{1}{2}$

   c. $-2 = \frac{2x+14}{x-1}$

4. Describe how each function was transformed from the parent function $f(x) = \frac{1}{x}$.

   a. $g(x) = \frac{1}{x-5} + 2$

   b. $h(x) = \frac{4}{x+5}$

   c. $j(x) = 2 - \frac{3}{x-1}$

5. As the rational function $y = \frac{1}{x}$ is translated, its asymptotes are also translated. Write an equation for the translation of $y = \frac{1}{x}$ that has the asymptotes described.

   a. horizontal asymptote $y = 0$ and vertical asymptote $x = 2$.

   b. horizontal asymptote $y = 2$ and vertical asymptote $x = -4$.

   c. horizontal asymptote $y = -4$ and vertical asymptote $x = 3$.

Discovering Advanced Algebra More Practice Your Skills

Graphs of Rational Functions

Name _____ Period _____ Date _____

1. Rewrite each rational expression in factored form.

   a. $\dfrac{x^2 - 5x - 6}{x^2 - 25}$

   b. $\dfrac{x^2 - 16}{6x^2 - 7x - 3}$

   c. $\dfrac{9x^2 - 1}{2x^3 - x^2 - 3x}$

2. Rewrite each expression in fractional form (as the quotient of two polynomials).

   a. $\dfrac{2}{x} + 3$

   b. $4 + \dfrac{2x - 7}{x + 5}$

   c. $\dfrac{5x - 7}{x + 3} - 4$

3. Find all vertical and horizontal asymptotes of the graph of each rational function.

   a. $f(x) = -\dfrac{1}{x^2}$

   b. $f(x) = \dfrac{3}{(x - 2)^2}$

   c. $f(x) = \dfrac{x^2 + x + 1}{x^2 - 4}$

4. Find all vertical asymptotes of the graph of each rational function.

   a. $f(x) = \dfrac{x^2 + 1}{x}$

   b. $f(x) = \dfrac{x^3}{x^2 - 4}$

   c. $f(x) = \dfrac{9 - x^2}{2 + x}$

5. Give the coordinates of all holes in the graph of each rational function.

   a. $f(x) = \dfrac{x - 3}{3 - x}$

   b. $f(x) = \dfrac{2x + 6}{x + 3}$

   c. $f(x) = \dfrac{x^2 - 3x - 10}{x + 2}$

Adding and Subtracting Rational Expressions

Name _____ Period _____ Date _____

1. Factor the numerator and denominator of each expression completely and reduce common factors.

   a. $\dfrac{x^2 - 4x}{x^2 - x - 12}$

   b. $\dfrac{x^2 - 49}{x^2 + 14x + 49}$

   c. $\dfrac{2x^2 - 10x}{3x^2 - 11x - 20}$

   d. $\dfrac{4x^2 - 1}{6x^2 - x - 2}$

   e. $\dfrac{9x^2 - 30x + 25}{9x^2 - 12x - 5}$

   f. $\dfrac{4x^2 + 21x + 5}{5x^2 + 23x - 10}$

2. Find the least common denominator for each pair of rational expressions. Don't forget to factor the denominators completely first.

   a. $\dfrac{3}{(x + 4)(x + 2)}, \dfrac{5}{(x + 4)(x - 5)}$

   b. $\dfrac{3x}{x^2 - 16}, \dfrac{2x}{x^2 + 5x + 4}$

   c. $\dfrac{2x - 1}{x^2 - 4x + 4}, \dfrac{3x^2}{x^2 - 6x + 8}$

   d. $\dfrac{x + 3}{x^2 - 7x - 8}, \dfrac{x - 5}{2x^2 + x}$

3. Add or subtract as indicated. Simplify by dividing any common factors.

   a. $\dfrac{3}{(x + 2)(x - 1)} + \dfrac{5}{(x + 1)(x - 1)}$

   b. $\dfrac{4}{x^2 - 49} - \dfrac{x}{(x + 7)(x - 1)}$

   c. $\dfrac{6}{x - 2} + \dfrac{x + 3}{2 - x}$

   d. $\dfrac{2}{x^2 - 5x + 4} + \dfrac{-2}{x^2 - 4}$

   e. $\dfrac{3x - 2}{x^2 + 2x - 24} - \dfrac{x + 3}{x^2 - 16}$

   f. $\dfrac{1}{x + 1} - \dfrac{x}{x - 2} + \dfrac{x^2 + 2}{x^2 - x - 2}$

Discovering Advanced Algebra More Practice Your Skills

Multiplying and Dividing Rational Expressions

Name _____ Period _____ Date _____

1. Use the commutative property of multiplication to rewrite the expressions and simplify.

   a. $\dfrac{(x-1)}{(x+9)} \cdot \dfrac{(x+9)}{(x+1)}$

   b. $\dfrac{(x-8)}{x^2} \cdot \dfrac{(x+11)}{x} \cdot \dfrac{x^3}{(x+11)}$

2. What is the domain of each expression in Exercise 1?

3. Multiply or divide as indicated. Simplify by dividing any common factors.

   a. $\dfrac{x^2+2x-15}{2x^2+9x-5} \cdot \dfrac{4x^2-1}{2x^2-5x-3}$

   b. $\dfrac{9x^2+6x}{2x^2-1} \div \dfrac{6x^2+x-2}{4x^2-4x+1}$

   c. $\dfrac{x^2-2x-35}{2x^3-3x^2} \cdot \dfrac{4x^3-9x}{7x-49}$

   d. $\dfrac{12x^2-22x+8}{3x} \div \dfrac{3x^2+2x-8}{2x^2+4x}$

   e. $\dfrac{6-2x}{x^2+4x+4} \cdot \dfrac{x^3+2x^2}{x^8-9x^6}$

   f. $\dfrac{x^2-36}{x^2-8x+16} \div \dfrac{3x-18}{x^2-x-12}$

4. Rewrite each fraction as a single rational expression.

   a. $\dfrac{\frac{2x-1}{x^2}}{\frac{2x^2+3x-2}{x}}$

   b. $\dfrac{\frac{x^2-9}{x^2-2x-3}}{\frac{x^2+6x+9}{x+1}}$

   c. $\dfrac{\frac{1}{x-2}+\frac{1}{x+2}}{\frac{x}{x+2}-\frac{x}{x-2}}$

Solving Rational Equations

Name _____ Period _____ Date _____

1. What values are excluded from the domains of the rational expressions in these equations?

a. $\dfrac{x}{x-2} + \dfrac{2}{3x} = \dfrac{2}{x-3}$

b. $\dfrac{5}{x^2-3x+2} - \dfrac{1}{x-2} = \dfrac{1}{3x-3}$

c. $\dfrac{x-5}{x^2-2x-8} = \dfrac{5x-1}{x^2-4}$

2. Solve.

a. $x - \dfrac{2}{x-3} = \dfrac{x-1}{3-x}$

b. $\dfrac{x}{x-5} + \dfrac{3}{x+2} = \dfrac{7x}{x^2-3x-10}$

c. $\dfrac{x}{x^2-1} + \dfrac{2}{x+1} = \dfrac{1}{2x-2}$

d. $\dfrac{x-1}{x-2} - \dfrac{2x-2}{x^2+3x} = \dfrac{5x-5}{x^2+x-6}$

e. $\dfrac{x+3}{x-3} + \dfrac{x}{x-5} = \dfrac{x+5}{x-5}$

f. $\dfrac{1}{2x} = \dfrac{5x+15}{x^2-6x} - \dfrac{x+6}{2x^2-12x}$

g. $\dfrac{x^2-3x-4}{x^3-x^2} - \dfrac{1}{x^2} = \dfrac{x-2}{x^2}$

h. $\dfrac{x^2+3x-18}{2x+10} = \dfrac{x-3}{2x+10} + 2x - 12$

Discovering Advanced Algebra More Practice Your Skills

Trigonometry and Trigonometric Functions

Right Triangle Trigonometry

Name _____ Period _____ Date _____

1. For each of the following right triangles, find the values of $\sin A$, $\cos A$, $\tan A$, $\sin B$, $\cos B$, and $\tan B$. (Write your answers as fractions in lowest terms.)

   a. 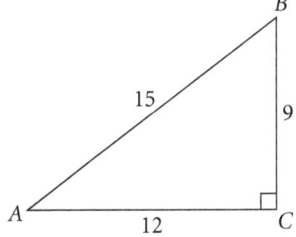   b.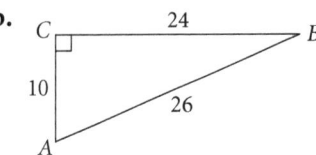

2. Draw a right triangle for each problem. Label the sides and angles with the given measures, then solve to find the unknown value. Round your answers to the nearest tenth.

   a. $\cos 27° = \dfrac{r}{8.5}$ b. $\tan^{-1}\left(\dfrac{9}{10}\right) = S$ c. $\cos 52° = \dfrac{z-3}{z}$

3. For each triangle, write an equation to calculate the labeled measure. Then, solve the equation.

   a. 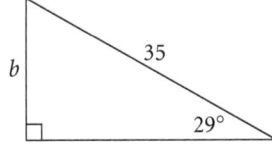   b. 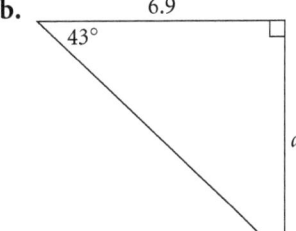   c.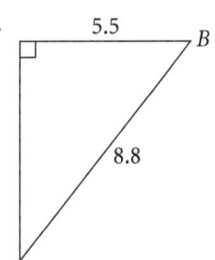

Discovering Advanced Algebra More Practice Your Skills

Extending Trigonometry

Name _____ Period _____ Date _____

1. Sketch each angle in the coordinate plane. Then, find the measure of the reference angle for each angle.

   a. 150° b. −30° c. 200°

2. Find the trigonometric value requested for each angle.

   a. cos θ

   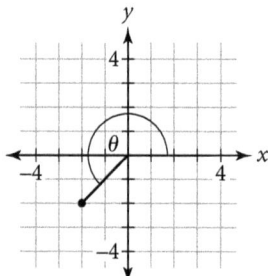

   b. tan α

   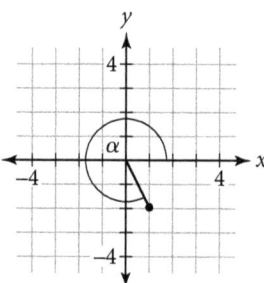

3. Without using a calculator, determine whether each value is positive or negative.

   a. tan 100° b. sin 195° c. cos 70°

   d. sin −100° e. tan 260° f. cos 260°

Defining the Circular Functions

Name _____ Period _____ Date _____

1. Find the exact value of each expression.

   a. $\cos 45°$ b. $\sin(-30°)$ c. $\cos 240°$ d. $\sin 360°$

2. Use your calculator to find each value, approximated to four decimal places. Then draw a diagram in a unit circle to represent the value. Name each reference angle.

   a. $\sin 37°$ b. $\cos 115°$ c. $\sin(-21°)$

3. Determine whether each function whose graph is shown below is periodic or not periodic. For each periodic function, identify the period.

   a.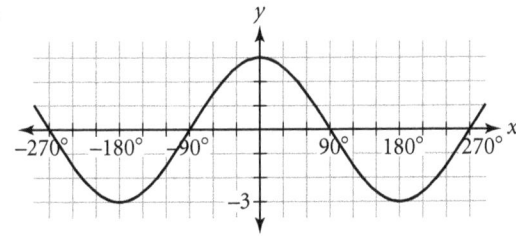

   b.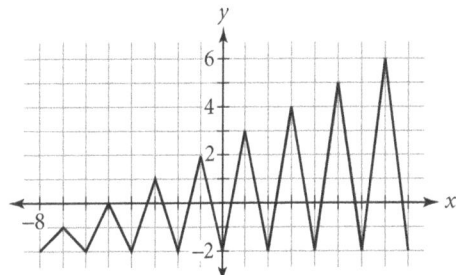

4. Identify an angle θ that is coterminal with the given angle. Use domain $0° \leq \theta \leq 360°$.

   a. $-42°$ b. $415°$ c. $913°$ d. $-294°$

5. Let θ represent the angle between the x-axis and the ray with endpoint (0, 0) passing through $(-3, 3)$. Find $\sin \theta$ and $\cos \theta$.

Radian Measure

Name _____ Period _____ Date _____

1. Convert between radians and degrees. Give exact answers.

    a. $\dfrac{5\pi}{4}$
    b. $15°$
    c. $330°$
    d. $-\dfrac{2\pi}{3}$

    e. $-140°$
    f. $780°$
    g. $-\dfrac{11\pi}{6}$
    h. $\dfrac{17\pi}{15}$

2. Find the length of the intercepted arc for each central angle.

    a. $r = 8$ and $\theta = \dfrac{5\pi}{4}$
    b. $r = 5.4$ and $\theta = 2.5$
    c. $d = 3$ and $\theta = \dfrac{\pi}{12}$

3. Solve for θ.

    a. $\sin\theta = \dfrac{\sqrt{3}}{2}$ and $90° \leq \theta \leq 180°$
    b. $\sin\theta = -1$ and $0° \leq \theta \leq 360°$

    c. $\cos\theta = \dfrac{1}{2}$ and $\pi \leq \theta \leq 2\pi$
    d. $\dfrac{\sin\theta}{\cos\theta} = \dfrac{1}{\sqrt{3}}$ and $0 \leq \theta \leq \dfrac{\pi}{2}$

4. The minute hand on a watch is 85 mm long. Round your answers in 4a and b to the nearest tenth, and in 4c to the nearest thousandth.

    a. What is the distance the tip of the minute hand travels, in mm?

    b. At what speed is the tip moving, in mm/min?

    c. What is the angular speed of the tip, in radians/min?

Graphing Trigonometric Functions

Name _____   Period _____   Date _____

1. Write an equation for each sinusoid as a transformation of the graph of either $y = \sin x$ or $y = \cos x$. More than one answer is possible. Describe the amplitude, period, phase shift, and vertical shift of each graph.

   a.
   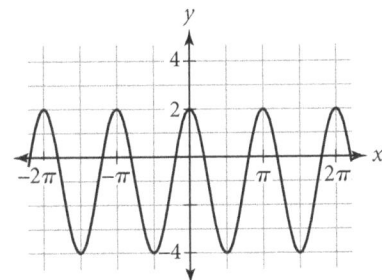

   b.
   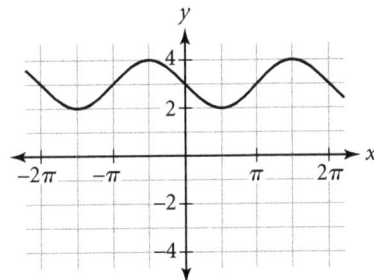

2. Graph each function for the interval $0 \leq \theta \leq 2\pi$.

   a. $y = 2\sin x + 1$

   b. $y = -\tan x$

   c. $y = 3\cos\left(x + \dfrac{\pi}{4}\right)$

   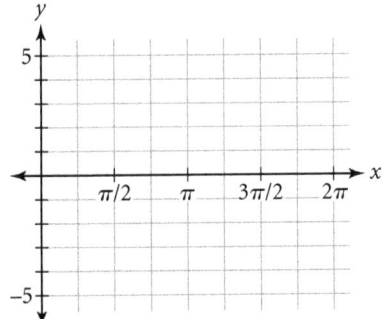
   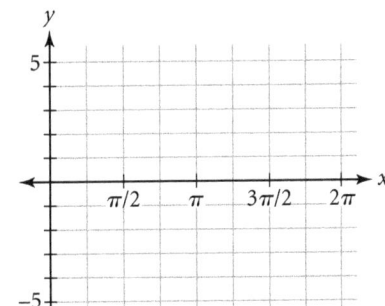
   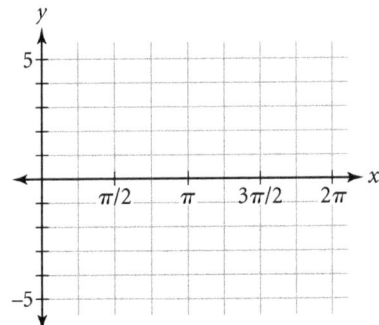

3. Write an equation for each sinusoid with the given characteristics.

   a. A cosine curve with amplitude 2.5, period 2π, and phase shift $\dfrac{\pi}{4}$

   b. A sine function with minimum value 2, maximum value 8, and one cycle starting at $x = 0$ and ending at $x = \dfrac{3\pi}{2}$

Discovering Advanced Algebra More Practice Your Skills
©2017 Kendall Hunt Publishing

Modeling with Trigonometric Equations

Name _____ Period _____ Date _____

1. Find all solutions for $0 \leq x < 2\pi$. Give exact values in radians.

  a. $\sin x = 1$

  b. $\cos x = 0$

  c. $\cos 3x = 0.5$

  d. $2\sin\left(\frac{1}{2}x\right) = 1$

  e. $\sin\frac{x}{2} = \frac{\sqrt{2}}{2}$

  f. $\cos\left(x - \frac{\pi}{4}\right) = \frac{\sqrt{3}}{2}$

2. Find all solutions for $0 \leq x < 2\pi$, rounded to the nearest hundredth.

  a. $3\cos(x + 0.4) = 2.6$

  b. $5 + 0.5\sin 2x = 4.7$

3. Consider the graph of the function
$$h = 8.5 + 5\sin\left[\frac{2\pi(t - 4)}{7}\right]$$

  a. What is the vertical translation?

  b. What is the average value?

  c. What is the vertical scale factor?

  d. What is the minimum value?

  e. What is the maximum value?

  f. What is the amplitude?

  g. What is the horizontal scale factor?

  h. What is the period?

  i. What is the horizontal translation?

  j. What is the phase shift?

4. The number of hours of daylight on any day of the year in Philadelphia, Pennsylvania, is modeled using the equation
$$y = 12 + 2.4\sin\left[\frac{2\pi(x - 80)}{365}\right]$$
where x represents the day number (with January 1 as day 1). This equation assumes a 365-day year (not a leap year).

  a. Find the number of hours of daylight in Philadelphia on day 172, the longest day of the year (the summer solstice).

  b. Find the day numbers of the two days when the number of hours of daylight is closest to 13.

Pythagorean Identities

Name _____ Period _____ Date _____

1. Evaluate. Give exact values.

   a. $\tan \dfrac{\pi}{3}$
   b. $\cot \dfrac{5\pi}{6}$
   c. $\sec \dfrac{\pi}{4}$

   d. $\csc \dfrac{4\pi}{3}$
   e. $\cot \pi$
   f. $\csc \dfrac{7\pi}{6}$

2. Find another function that has the same graph as each function below. (More than one answer is possible.)

   a. $y = \tan(x + \pi)$
   b. $y = \sin(x - 2\pi)$
   c. $y = -\csc(x - 2\pi)$

3. Use trigonometric identities to rewrite each expression in a simplified form containing only sines and cosines, or as a single number.

   a. $\tan \theta + \sec \theta$
   b. $(\sec^2 \theta - \tan^2 \theta)\cos^2 \theta$

   c. $\cot \theta \sin^2 \theta - \tan \theta \cos^2 \theta$
   d. $(\csc \theta + \cot \theta)(\csc \theta - \cot \theta)$

4. Determine whether each equation is an identity or not an identity.

   a. $\sin(A + \pi) = \cos A$
   b. $\tan\left(A - \dfrac{\pi}{2}\right) = -\cot A$

   c. $\csc^2 A = \cot A(\tan A + \cot A)$
   d. $\sec A \cot A = \csc A$

Discovering Advanced Algebra More Practice Your Skills

Probability

Randomness and Probability

Name _____ Period _____ Date _____

1. A national survey was taken measuring the highest level of educational achievement among adults age 30 and over. Express each probability to the nearest 0.001.

Highest level of education	Women	Men	Total
8th grade or less	35	46	81
High school graduate	232	305	537
Some college	419	374	793
Bachelor's degree	539	463	1002
Graduate or professional degree	377	382	759
Total	1602	1570	3172

   a. What is the probability that a randomly chosen person from the survey group is a man?

   b. What is the probability that the highest level of education completed by a randomly chosen person from the survey group is a bachelor's degree?

   c. What is the probability that a randomly chosen woman has earned a bachelor's or graduate degree?

   d. Are the probabilities in 1a–c experimental or theoretical?

2. A ping pong ball is dropped into a flat-bottomed box with this target. What is the probability the ball stops in the center?

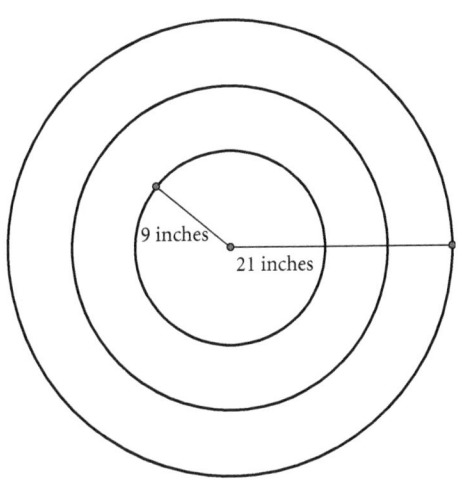

Multiplication Rules of Probability

Name _____ Period _____ Date _____

1. Find the probability of each branch or path, a–g, in the tree diagram below.

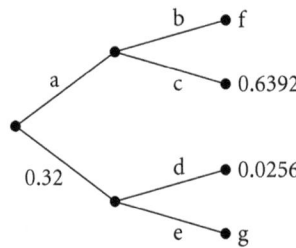

2. Draw a tree diagram that shows all possible equally likely outcomes if a penny is tossed once and then a six-sided die is rolled once. Then use your diagram to find each probability.

   a. What is the probability of tossing a head and rolling a 5?

   b. What is the probability of tossing a tail and rolling an even number?

3. Liam draws one playing card from a 52-card deck and places it on the table. Then he draws a second card and places it to the right of the first card. What is the probability that both cards are black?

4. Three friends try out for sports teams at their high school. Gladys tries out for the lacrosse team and has a 40% chance of success (making the team). Becky tries out for the swim team and has a 30% chance of success. Serita tries out for the tennis team and has a 25% chance of success. Use the tree diagram to find each probability.

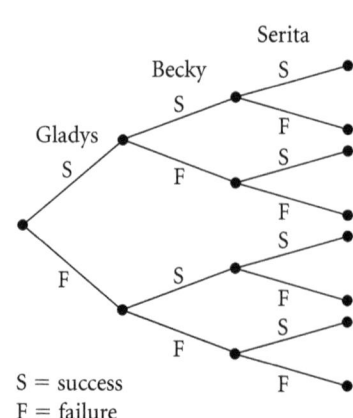

   a. What is the probability that all three girls will make their teams?

   b. What is the probability that exactly one of the girls will be successful?

Addition Rules of Probability

Name _____ Period _____ Date _____

1. Refer to the Venn diagram, which gives probabilities related to the two events "plays the piano" and "plays the violin." These probabilities apply to the students at Riverway Middle School, which has 800 students.

   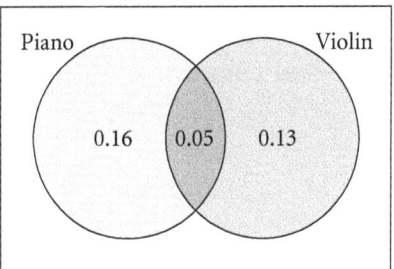

   a. What is the probability that a randomly chosen student at Riverway plays the piano?

   b. How many students at Riverway play both instruments?

2. Refer to the Venn diagram in Exercise 1. Find the probability of each event indicated. N represents the event that a student plays the piano, and V represents the event that a student plays the violin.

   a. $P(\text{not } N)$ b. $P(N \text{ or } V)$ c. $P(\text{not } N \text{ and not } V)$

3. For a class project, Diana surveys 300 students at her high school about the entertainment equipment (DVR, gaming systems, and tablets) they have in their homes. She gathers the following information.

   187 homes had gaming systems and 141 homes had tablets.

   19 homes had no entertainment equipment, whereas 12 homes had tablets only.

   81 homes had gaming systems and DVR, but not tablets.

   11 homes had gaming systems and tablets, but not DVR.

   43 homes had DVR and tablets, but not gaming systems.

   a. Complete the Venn diagram using probabilities.

   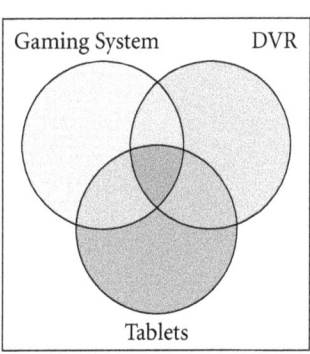

   b. What is the probability that a student's home has a DVR, but neither a gaming system nor a tablet?

   c. What is the probability that a student's home has all three pieces of equipment?

Discovering Advanced Algebra More Practice Your Skills

Bivariant Independence

Name _____ Period _____ Date _____

1. Find the marginal totals for each row and column.

Vacation at the Beach	Vacation in the Mountains	
	Yes	No
Yes	58	72
No	62	138

2. Given the two-way table in problem 1, find the requested conditional proportions.

   a. Proportion who vacation in the mountains given they vacation at the beach.

   b. Proportion who vacation in the mountains given they do not vacation at the beach.

   c. Proportion who vacation at the beach given they vacation in the mountains.

   d. Proportion who vacation at the beach given they do not vacation in the mountains.

3. Create a segmented bar plot for …

   a. Proportions based on the conditions of vacationing at the beach.

   b. Proportions based on the conditions of vacationing in the mountains.

4. Are vacationing at the beach and vacationing in the mountains independent of each other? How do you know?

5. Find the sample proportion and estimate the true proportion given …

   a. $x = 30, n = 110$

   b. $x = 54, n = 236$

   c. $x = 198, n = 992$

   d. $x = 382, n = 1958$

Applications of Statistics

Experimental Design

Name _____ Period _____ Date _____

For Exercises 1 and 2, use each of the three scenarios to answer the questions.

   a. The Glass Pane Company calls numbers on a Monday that were randomly chosen from the telephone book and asks the person who answers the phone, "In what year were the windows in your home installed?"

   b. Mario wanted to determine whether temperature affects the level of animal activity. At a park near his house one day he recorded the number of animals he saw between 1:00 P.M. and 2:00 P.M. when it was 85°F. Then at the same park on another day he recorded the number of animals he saw between 1:00 P.M. and 2:00 P.M. when it was 55°F.

   c. Sienna wants to know if the citizens of her town think the price of a movie ticket is fair and asks every other person standing in the ticket line, "Do you think movie tickets are overpriced?"

1. Identify which type of data collection was used.

2. Identify at least one source of bias in each study design in each scenario. Explain your reasoning.

3. A radio station announce asked listeners to call in and four of the five listeners who responded were under 20 years old. She then stated that over 75% of her listeners are under the age of 20. Is this statistic reliable? Explain.

4. Scott thinks eating fruit before a test will increase a person's score on the test. Just before a test in his class, he randomly gives fruit to half of his classmates, and those students eat the fruit. The rest of the students do not eat fruit. He then compares the scores after the test.

   a. Identify which type of data collection was used.

   b. Identify the treatment and state how it was assigned.

Normal Distributions

Name _____ Period _____ Date _____

1. Use the graphs to estimate the mean and standard deviation of each distribution.

   a.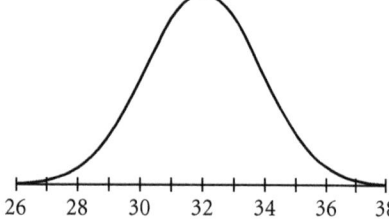

   b.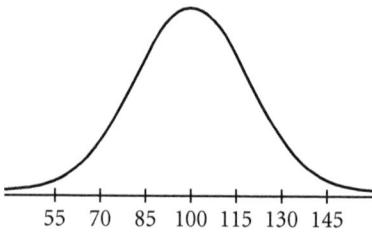

2. Estimate the equation of each graph in Exercise 1.

3. From each equation, find or estimate the mean and standard deviation.

   a. $y = \dfrac{1}{8\sqrt{2\pi}} \left(\sqrt{e}\right)^{-[(x-55)/8]^2}$

   b. $y = \dfrac{0.4}{0.75}(0.60653)^{-[(x-4.8)/0.75]^2}$

4. The weights of 1000 children were recorded on their first birthdays. The weights are normally distributed with mean 10.3 kg and standard deviation 1.6 kg.

   a. What is the probability that a randomly selected child will weigh less than 8.7 kg?

   b. What is the probability that a randomly selected child will weigh between 8.7 kg and 11.9 kg?

   c. How many of the 1000 children would you expect to weigh between 7.1 kg and 13.5 kg?

   d. How many of the 1000 children would you expect to weigh more than 8.7 kg?

z-Values and Confidence Intervals

Name _____ Period _____ Date _____

1. The heights of a group of 500 women are normally distributed with mean 65 inches and standard deviation 2.2 inches. Find the height for each of these z-values to the nearest tenth of an inch.

   a. $z = 2$ b. $z = 0.5$ c. $z = -3.4$

2. The mean commuting time for a resident of a certain metropolitan area is 38 minutes, with a standard deviation of 10 minutes. Assume that commuting times for this area are normally distributed.

   a. Find the z-value for a 23-minute commute.

   b. Find the z-value for a 60-minute commute.

   c. What is the probability that a commute for a randomly chosen resident will be between 28 minutes and 58 minutes?

3. A sample has mean 52.6 and standard deviation 6.4. Find each confidence interval. Assume $n = 25$ in each case.

   a. 68% confidence interval

   b. 95% confidence interval

   c. 99.5% confidence interval

4. Given the population set of {1, 3, 4, 7}.

   a. Find mean and standard deviation of this data.

   b. Find the 95% margin of error in estimating the true mean.

   c. Give a 95% confidence interval for the mean.

Discovering Advanced Algebra More Practice Your Skills
©2017 Kendall Hunt Publishing

Bivariate Data and Correlation

Name _____ Period _____ Date _____

1. Complete this table. Then answer 1a–d to calculate the correlation coefficient.

x	y	$x - \bar{x}$	$y - \bar{y}$	$(x - \bar{x})(y - \bar{y})$
2	5			
4	8			
6	10			
8	13			
10	15			
12	18			

   a. What are $\bar{x}$ and $\bar{y}$?

   b. What is the sum of the values for $(x - \bar{x})(y - \bar{y})$?

   c. What are s_x and s_y?

   d. Calculate $r = \dfrac{\sum (x - \bar{x})(y - \bar{y})}{s_x s_y (n - 1)}$.

   e. What does this value of r tell you about the data?

2. For each research finding, decide whether the relationship is causation, correlation, or both. If it is only correlation, name a possible lurking variable that may be the cause of the results.

   a. A pharmaceutical company made a television commercial based on the finding that people who took their daily vitamins lived longer. Do vitamins extend life?

   b. A high school counselor observed that students who take a psychology course are never in the school band. Does this mean that students who are interested in psychology have less musical ability than other students do?

www.ingramcontent.com/pod-product-compliance
Ingram Content Group UK Ltd.
Pitfield, Milton Keynes, MK11 3LW, UK
UKHW051537180825
7456UKWH00024B/227